Synthetic Biology Conference (SEED 2017)

Engineering, Evolution, and Design

Vancouver, Canada
20-23 June 2017

ISBN: 978-1-5108-5673-8

Printed from e-media with permission by:

Curran Associates, Inc.
57 Morehouse Lane
Red Hook, NY 12571

Some format issues inherent in the e-media version may also appear in this print version.

Printed by Curran Associates, Inc. (2018)

AIChE
120 Wall Street, FL 23
New York, NY 10005-4020

Phone: (800) 242-4363
Fax: (203) 775-5177

www.aiche.org

Additional copies of this publication are available from:

Curran Associates, Inc.
57 Morehouse Lane
Red Hook, NY 12571 USA
Phone: 845-758-0400
Fax: 845-758-2633
Email: curran@proceedings.com
Web: www.proceedings.com

Page:

- Weston Kightlinger
- Liang Lin
- Milan Mrksich
- Michael C. Jewett

2:45 PM
12 Expanding the Portfolio of Bio-Manufacturing: Advanced Automation, Biology, and Computational Tools for Rapid Microbe Engineering

- Victoria Hsiao

3:00 PM
13 Any Molecule in Any Microbe, from Lab to Full-Scale Manufacturing

- Kasia Gora

3:15 PM
14 Vmax™ – a Novel Microbial Host for the Biotech Industry

- Matthew Weinstock

3:30 PM
15 You Design it, We Build it

- Emily Leproust

4:15 PM
16 A Butyrate-Producing Synthetic Biotic to Treat Inflammatory Bowel Disease In Vivo

- Adam Fisher
- Vincent M. Isabella
- Paul Miller

4:45 PM
17 Engineering the Mammalian Gut Microbiome in Situ

- Harris H. Wang
- Carlotta Ronda
- Sway Chen
- Vitor Cabral

5:15 PM
18 Paper-Based Diagnostic for Assessing the Gut Microbiome

- Melissa K. Takahashi
- Xiao Tan

4:30 PM

30 Universal Chimeric Antigen Receptors for Multiplexed and Logic Control of T Cell Responses

- Wilson Wong

5:00 PM

31 Programming Protein Function with Synthetic RNA Splicing Devices

- Melina Mathur
- Cameron Kim
- Sarah Munro
- Christina D. Smolke

5:15 PM

33 Understanding and Mitigating the Effects of Limited Transcriptional Resources in Mammalian Cells

- Yili Qian
- Ross Jones
- Ron Weiss
- Domitilla Del Vecchio

8:30 AM

35 Harnessing the Chemistry of Plant Natural Product Biosynthesis

- Sarah O'Connor

10:00 AM

36 Conversion of Glycerol to 3-Hydroxypropanoic Acid By Genetically Engineered Bacillus Subtilis

- Aida Kalantari
- Boyang Ji
- Tao Chen
- Ivan Mijakovic

10:30 AM

37 Improving E. coli Methyl Ketone Production during Fed-Batch Fermentation By Co-Factor Balancing

- Ee-Been Goh
- Edward E. K. Baidoo
- David E. Ando
- Jay D. Keasling
- Hector Garcia Martin

- Harry R. Beller

11:00 AM

39 Production of Industrial Products from Natural Gas Using an Engineered Methanotroph Biocatalyst

- Mark Held
- James Kealey

1:30 PM

40 Directed Evolution of Targeted Cas9 to Specific Cleavage Sites

- Barrett Steinberg
- Derek Cerchione
- Morgan Maeder
- Maxwell Skor
- Hari Jayaram
- Vic Myer

2:00 PM

41 CRISPR Technologies 2.0: Harnessing the Next Generation of Cas Nucleases for Genome Editing and Beyond

- Chase L. Beisel

2:30 PM

43 Rational Engineering of Chimaeric Recombinases for Genomic Excision

- Jumai A. Abioye
- W. Marshall Stark
- Femi J. Olorunniji
- Julien Reboud
- Jonathan M. Cooper
- Stephanie E. Holt
- Chris M. Proudfoot

3:45 PM

45 Twin-Primer Non-Enzymatic DNA Assembly: An Efficient and Accurate Multi-Part DNA Assembly Method

- Jing Liang
- Zihe Liu
- Ee Lui Ang
- Huimin Zhao

4:15 PM

- Murat Arcak
- Marcella M. Gomez

11:30 AM

58 A Synthetic Biology Workflow Using Sbol

- Chris J. Myers
- James McLaughlin
- Goksel Misirli
- Tramy Nguyen
- Meher Samineni
- Michael Zhang
- Zach Zundel
- Anil Wipat

POSTERS

60 A Comprehensive Characterization of the Nitroreductase Superfamily

- David Morales
- Janine Copp
- Eyal Akiva
- Patsy Babbitt
- Nobu Tokuriki

62 A CRISPR/Cas9 Genome Engineering Toolbox for Systems Metabolic Engineering of Yeast

- Rene Verwaal
- Nathalie Buiting-Wiessenhaan
- Brenda Vonk
- Johannes A. Roubos

63 A Genetic Switch for Stable, Long-Term Fermentative Production of Anabolic Products in Yeast

- Penelope Chua
- Adam Meadows
- Hanxiao Jiang

64 A Massively Parallel Assay for Studying 5' UTR Dependent Translation Regulation in Mammalian Cells

- Georg Seelig
- Paul Sample
- Ban Wang

302 Using Yeast to Identify Peptides That Regulate Human Oncogenic Transcription Factors

- Chi-Fang Wu
- John Koschwanez

303 Vectoreditorjs: An Open-Source Web-Based Javascript Application for DNA Sequence Visualization, Annotation, and Editing

- Sarah LaFrance
- Leanne Springthorpe
- Hector Plahar
- Jacob Coble
- Thomas Rich
- Nathan J Hillson
- Joshua Nixon

305 Why Doesn't Everyone Do Everything? the Cooperation Conundrum of Engineered Consortia for the Catabolism of Phosphotriester Pesticides

- Dan Kehila
- Nobu Tokuriki

*****The papers in this book may contain font errors that were present in the publisher provided material.*****

Synthetic Epigenetics

- Conference: Synthetic Biology Engineering Evolution Design SEED
- Year: 2017
- Proceeding: 2017 Synthetic Biology: Engineering, Evolution & Design (SEED)
- Group: General Submissions
- Session:

  Session 1: Molecular Programming

- Time:

  Tuesday, June 20, 2017 - 10:30am-11:00am

Authors:
Khalil, A. S., Boston University
Cells with identical genomes can exhibit distinct phenotypes and states that are stably maintained. This capacity is achieved with self-propagating systems that maintain molecular memory of past signals. I will discuss our efforts to apply synthetic biology to explore and engineer these biological systems. Specifically, I will discuss the development of synthetic tools and circuits for controlling: (1) molecular changes to chromatin to produce gene expression programs and memory; and (2) an altogether different form of epigenetics, encoded in the self-propagating conformations of prion proteins. The predictive engineering of these epigenetic systems will expand the types of cellular responses that can be achieved, and enable memories and traits to be encoded into synthetic biological organisms.

Controlling Materials with Biomolecular Circuits

- Conference: Synthetic Biology Engineering Evolution Design SEED
- Year: 2017
- Proceeding: 2017 Synthetic Biology: Engineering, Evolution & Design (SEED)
- Group: General Submissions
- Session:

  Session 1: Molecular Programming

- Time:

  Tuesday, June 20, 2017 - 11:00am-11:30am

Authors:
Schulman, R., Johns Hopkins University
Cells can sense chemical signals in the environment and in response release or degrade different molecular species and change their shape, size or mechanical properties. Complex chemical programs implemented by biomolecular circuits allow cells to have complex, many stage responses to a single stimulus. Molecular programming techniques could enable materials to likewise be controlled by biomolecular circuits and execute different complex adaptive changes in response to different stimuli. Using hydrogels as a model, we explore how molecular circuits could interpret molecular signals and direct complex responses in abiotic materials.

Towards Developing Complex, Environmentally-Responsive Genetic Circuits for Real-World Applications

- Conference: Synthetic Biology Engineering Evolution Design SEED
- Year: 2017
- Proceeding: 2017 Synthetic Biology: Engineering, Evolution & Design (SEED)
- Group: General Submissions
- Session:

  Session 1: Molecular Programming

- Time:

  Tuesday, June 20, 2017 - 11:30am-12:00pm

Authors:
Moon, T. S., Washington University in St. Louis
Shopera, T., Washington University in St. Louis
Hoynes-O'Connor, A., Washington University in St. Louis
Lee, Y. J., Washington University in St. Louis
Immethun, C., Washington University in St. Louis

As synthetic biology transitions out of the lab and towards real-world applications, there is a need to develop a well-characterized toolbox of genetic sensors and regulators that can respond to extrinsic environmental conditions, instead of lab chemicals (e.g., IPTG). In addition, many sophisticated cellular functions require complex genetic circuits, which can lead to significant consumption of finite cellular resources and thus resource-coupled interference in predictable circuit behavior. To solve these problems, we have developed environmental sensors, RNA regulators, and feedback circuits. First, in multiple bacterial species, we have built and characterized sensors that respond to pH, temperature, light, oxygen levels, nitrogen source levels, or concentrations of diverse compounds found in a variety of environments. These environmental sensors have also been combined to construct multi-input logic gates (e.g., AND gates). Second, we have employed the CRISPR interference system (CRISPRi) and synthetic antisense RNAs (asRNAs) to repress or derepress target genes in a programmable manner. Importantly, we provided design rules for asRNA regulators, based on a comprehensive study of more than a hundred *de novo* designed asRNA regulators, enabling predictable implementation of these regulators in complex genetic circuits (e.g., NAND gates that respond to pH and temperature). Third, we have investigated the effects of resource competition on circuit behaviors. For the first time, we experimentally demonstrated that negative feedback can significantly reduce resource-coupled interference in complex circuits, allowing for robust circuit performance. These environmental sensors, RNA regulators, and feedback motif lay the groundwork for the implementation of complex, environmentally-responsive genetic circuits in metabolic engineering, environmental, and medical applications.

Construction and Characterization of Consensus Toggle Switches

- Conference: Synthetic Biology Engineering Evolution Design SEED
- Year: 2017
- Proceeding: 2017 Synthetic Biology: Engineering, Evolution & Design (SEED)
- Group: General Submissions
- Session:

  Session 1: Molecular Programming

- Time:

  Tuesday, June 20, 2017 - 12:00pm-12:15pm

Authors:
de Freitas Magalhães, B., Rice University
Sontag, E., Rutgers-The State University of New Jersey
Bennett, M. R., Rice University
A genetic toggle switch is a circuit that possesses two stable opposite states of gene expression. Many synthetic circuits that are applied to biotechnology are based on toggle switches, such as biosensors. However, extrinsic and intrinsic noise can cause random switching and generate a variety of gene expression states. To make circuits more robust, we propose the creation of a consensus toggle switch that will have the ability to amplify induction signals among neighboring cells and achieve a consensus in the population. We will create the consensus by employing quorum sensing in our circuits. Quorum sensing (QS) is the mechanism through which bacteria communicate with each other, by producing signaling molecules that can affect gene expression in a density-dependent way. We hypothesize that synthetic toggles that coordinate their behavior with QS can better perform tasks that are difficult for uncoordinated systems. Here, we constructed 4 versions of consensus toggles to understand how the presence/absence of positive feedback loops in quorum-sensing genes and direct/indirect repression affect toggle behavior. In our designs, each possible state of the toggle is reported by one color: yellow (YFP) or blue (CFP). We find that the dynamics of consensus toggle switches depend on their regulatory topologies. Topologies that contain indirect repression show a peak of YFP expression followed by a shutdown, while topologies with direct repression do not shut down. Furthermore, topologies with indirect repression do not show distinguishable CFP expression between induced and uninduced samples, while topologies with direct repression do. Topologies that lack extra feedback loops show less variability, while topologies that contain these extra feedback loops are more variable. These results enable the selection of a topology that works better for the creation of a consensus toggle. In the microscope, the toggle with direct repression and without extra feedback loops showed

faster switch compared to a conventional toggle. These results are encouraging, because a consensus toggle has the potential of generating more sensitive and robust detection of signals than regular toggles. And to further apply this circuit, such as for the detection of diseases/infections in the mammalian gastrointestinal tract or for the surveillance and bioremediation of environments, it is desired that the circuit shows sensitivity and robustness.

Deep Learning of the Regulatory Grammar of Yeast 5' Untranslated Regions from 500,000 Random Sequences

- Conference: Synthetic Biology Engineering Evolution Design SEED
- Year: 2017
- Proceeding: 2017 Synthetic Biology: Engineering, Evolution & Design (SEED)
- Group: General Submissions
- Session:

  Session 2: High Throughput Design Space Exploration

- Time:

  Tuesday, June 20, 2017 - 1:45pm-2:15pm

Authors:
Groves, B., University of Washington
Our ability to predict protein expression from DNA sequence alone remains poor, reflecting our limited understanding of cis-regulatory grammar and hampering the design of engineered genes for synthetic biology applications. We have generated a model that predicts the translational efficiency of the 5â□™ untranslated region (UTR) of mRNAs in the yeast *Saccharomyces cerevisiae.* We constructed a library of nearly half a million 50 nucleotide-long random 5â□™ UTRs and assayed them in a single massively parallel growth selection experiment. The resulting data have allowed us to quantify the impact on translation of Kozak sequence composition, upstream open reading frames (uORFs) and secondary structure. With this data, we have trained a convolutional neural network on the random library and validated it by predicting the translational efficiency of 5â□™ UTRs that natively occur in yeast. The model additionally was used to computationally evolve highly translating 5â□™ UTRs. We have confirmed experimentally that the great majority of the evolved sequences lead to higher translation rates than the starting sequences, demonstrating the predictive utility of this model.

Turbocharge RNA Device Engineering with Data-Rich Strategies and Automation

- Conference: Synthetic Biology Engineering Evolution Design SEED
- Year: 2017
- Proceeding: 2017 Synthetic Biology: Engineering, Evolution & Design (SEED)
- Group: General Submissions
- Session:

  Session 2: High Throughput Design Space Exploration

- Time:

  Tuesday, June 20, 2017 - 2:15pm-2:30pm

Authors:
Xiang, J. S., Stanford University
Smolke, C. D., Stanford University
Townshend, B., Stanford University
Manzanarez, G., Stanford
Hinks, M. M., Stanford University
Engineered biological systems hold great potential to transform technologies in health, medicine and sustainability. Challenges in the design of biological devices include our incomplete understanding of the sequence-function relationships underlying the design space, and the limited throughput of experiments required to characterize individual devices. To significantly improve the efficiency of engineering highly functional gene-regulatory RNA devices, we 1) developed a massively parallel FACS-seq screen to assay the activity of hundreds of thousands of ribozyme switch library variants in yeast and mammalian cells, 2) adapted and automated a similarly high-throughput *in vitro* Cleave-seq assay on a liquid handler for parallelization and rapid iteration of the design-build-test cycle, and 3) employed computational analyses to extract design principles from the large datasets to inform future rational design.

To assay their activity in yeast or mammalian cells, ribozyme switches - a class of RNA devices with ligand-controllable self-cleavage activity - are integrated into the untranslated region of a fluorescent reporter, such that cleavage directly modulates fluorescence levels. Combining fluorescence-activated cell sorting (FACS) and next-generation sequencing (NGS), we extensively interrogated the effect of tertiary loop I-loop II interaction motifs on ribozyme activity, characterizing 10^5 variants in a single FACS-seq experiment[1]. Cleave-seq measures *in vitro* cleavage activities of the ribozyme switches at a similar throughput using NGS, and automation of this method allowed us to quickly iterate through different library designs, while obtaining insight into differences in conserved sequence motifs between *in vitro* experiments and inside cells. With these high throughput techniques, we generated new RNA switches to a number of small molecule therapeutics and metabolites, which can serve as gene-control devices and live cell genetic sensors. Our analysis of the large datasets also identified a set of general sequence motifs that result in enhanced ribozyme cleavage activity, highlighting the importance of pseudoknot-like base pairs at the base of the ribozyme stem loops to mediate tertiary interactions. These datasets serve as a rich resource to further enable the discovery of novel hammerhead ribozymes with bioinformatics approaches and streamline the design of RNA switches for synthetic biology applications.

[1]Townshend, B., Kennedy, A. B., Xiang, J. S., & Smolke, C. D. (2015). High-throughput cellular RNA device engineering. Nature methods, 12(10), 989-994.

Design of Protein Glycosylation Sites By Cell-Free Protein Synthesis and Mass Spectrometry of Self-Assembled Monolayers

- Conference: Synthetic Biology Engineering Evolution Design SEED
- Year: 2017
- Proceeding: 2017 Synthetic Biology: Engineering, Evolution & Design (SEED)
- Group: General Submissions
- Session:

  Session 2: High Throughput Design Space Exploration

- Time:

  Tuesday, June 20, 2017 - 2:30pm-2:45pm

Authors:
Kightlinger, W., Northwestern University
Lin, L., Northwestern University
Mrksich, M., Northwestern University
Jewett, M. C., Northwestern University
Protein glycosylation, the post-translational attachment of complex oligosaccharides (glycans), is the most abundant polypeptide modification in eukaryotes, and is integrally involved in human biology and disease. Changes in glycosylation have been shown to modulate the pharmacokinetics and potency of protein therapeutics and vaccines, making them critical quality attributes for biotechnology applications. However, the study of glycans in native systems (glycobiology) and the intentional manipulation of protein glycosylation patterns (glycoengineering) remain limited by a shortage of methods to characterize glycosylation enzyme activities and a reliance on endogenous glycosylation systems which can produce highly variable glycosylation patterns and constrain choices of production hosts, protein trafficking, and possible glycoforms. The structural complexity of enzymes involved in initial modification of polypeptides with glycans and the vast sequence spaces which must be surveyed to determine their specificities have impeded efforts to implement synthetic glycosylation systems in heterologous production hosts, such as *Escherichia coli*. New tools are required to rapidly express polypeptide modifying glycosyltransferases and determine design rules for engineering protein glycosylation sites which can be efficiently and specifically modified with a glycan of interest.

Here we describe an *in vitro* platform for high-throughput expression and characterization of glycosylation enzymes using *E. coli* Cell-Free Protein Synthesis (CFPS) and Self-Assembled Monolayers for Desorption Ionization Mass Spectrometry (SAMDI-MS). This workflow allowed us to produce >800 μg/mL of a cytoplasmic N-linked glycosyltransferase (NGT) *in vitro* and determine its peptide acceptor and sugar donor specificities at unprecedented depth and throughput with ~10,000 unique reactions conditions and ~3,000 unique peptide substrates. We used this dataset to develop a small, robust acceptor sequence motif (GlycTag) to direct the efficient installation of N-linked glycans onto the internal loops of heterologous protein substrates *in vitro* and in the cytoplasm of living *E. coli*. We found that SAMDI-MS observations at the peptide level accurately predicted modification trends in the context of whole proteins. We placed GlycTags in three structurally diverse target proteins, including the Fc region of a human antibody, and found that these sequences were modified at 3-5 fold greater efficiency compared to wild-type glycosylation sequence motifs. Notably, this N-linked glycosylation system does not require protein transport across cellular membranes or the use of membrane bound components, making it particularly attractive for the development of synthetic glycosylation systems.

This work provides a disruptive technique for glycosyltransferase characterization and provides a new set of tools towards emerging bacterial and *in vitro* glycosylation platforms. Ultimately, these glycoengineering efforts will enable a deeper understanding of glycan structure and function and help to bring about a new generation of rationally designed glycoprotein therapeutics, vaccines, and diagnostics.

Expanding the Portfolio of Bio-Manufacturing: Advanced Automation, Biology, and Computational Tools for Rapid Microbe Engineering

- Conference: Synthetic Biology Engineering Evolution Design SEED
- Year: 2017
- Proceeding: 2017 Synthetic Biology: Engineering, Evolution & Design (SEED)
- Group: General Submissions
- Session:

  Session 3: Manufacturing

- Time:

  Tuesday, June 20, 2017 - 2:45pm-3:00pm

Authors:
Hsiao, V., Amyris
Amyris is an end-to-end biological manufacturer that uses synthetic biology to engineer microbes as catalysts for conversion of sugar into chemical products. Many of these chemical targets were previously derived from limited natural resources and economically inaccessible via traditional chemistries. In the last decade, Amyris has developed a sophisticated technology platform for engineering microbes to produce high levels of isoprenoids including artemisinin and farnesene. In recent years, we have begun to harness the power of this proven infrastructure to expand the portfolio of bio-manufactured chemical products.

In this talk, I will describe the current state-of-the-art tools at Amyris, including our automated strain engineering, high-throughput screening, and fermentation. Then I will discuss how these tools are evolving under the DARPA-funded â□œMgs to Kgsâ□□ project to enable the concurrent development of 100â□™s of products simultaneously. In the first year, the project has produced more than 150 target molecules from all parts of metabolic space. As the project continues, â□œMgs to Kgsâ□□ will enable Amyris to incorporate next generation high-throughput â□˜omics, automation, data science, and machine learning tools to further increase the rate at which biology can be engineered.

Any Molecule in Any Microbe, from Lab to Full-Scale Manufacturing

- Conference: Synthetic Biology Engineering Evolution Design SEED
- Year: 2017
- Proceeding: 2017 Synthetic Biology: Engineering, Evolution & Design (SEED)
- Group: General Submissions
- Session:

  Session 3: Manufacturing

- Time:

  Tuesday, June 20, 2017 - 3:00pm-3:15pm

Authors:
Gora, K.

Identifying, understanding, and ultimately engineering the complex set of interacting genes responsible for any cellular phenotype is extremely difficult and traditionally requires years of study. For example, optimizing metabolic flux to a native or non-native small molecule or increasing the diversity of carbon sources that a microbe can use are well-studied topics, and yet very few novel compounds have made it through the high risk path to commercialization. To address this problem, Zymergen has developed a powerful platform for automated high-throughput microbial strain development that is focused on improving the economics of large-scale fermentation processes and accelerating the industrialization of novel bio-products. The platform has been designed to be flexible, highly reliable, and host-agnostic. With its high throughput capability Zymergen tests thousands of design ideas across many library types to identify those that improve performance. Automation makes Zymergenâ□™s strain improvement process both precise and repeatable. Using this platform, Zymergen takes a data-driven approach to biology, capturing as much data as possible and using heuristically-guided search processes that embrace the complexity of biology.

Vmax™ – a Novel Microbial Host for the Biotech Industry

- Conference: Synthetic Biology Engineering Evolution Design SEED
- Year: 2017
- Proceeding: 2017 Synthetic Biology: Engineering, Evolution & Design (SEED)
- Group: General Submissions
- Session:

  Session 3: Manufacturing

- Time:

  Tuesday, June 20, 2017 - 3:15pm-3:30pm

Authors:
Weinstock, M., Synthetic Genomics

This presentation will focus on Vmaxâ„¢, a high performing bacterium being developed at Synthetic Genomics, Inc. as a next-generation host organism for biotechnology and molecular biology applications[1].Â We will describe the current state of the platform, and will discuss how technology and genome design principles developed over the course of the minimal cell project[2] are being applied to engineer advanced Vmaxâ„¢ strains.

1. Weinstock, M. T., Hesek, E. D., Wilson, C. M. & Gibson, D. G. *Vibrio natriegens* as a fast-growing host for molecular biology. *Nat. Methods* **13,** 849â€"851 (2016).
2. Hutchison, C. A. *et al.* Design and synthesis of a minimal bacterial genome. *Science* **351,** aad6253 (2016).

You Design it, We Build it

- Conference: Synthetic Biology Engineering Evolution Design SEED
- Year: 2017
- Proceeding: 2017 Synthetic Biology: Engineering, Evolution & Design (SEED)
- Group: General Submissions
- Session:

  Session 3: Manufacturing

- Time:

  Tuesday, June 20, 2017 - 3:30pm-3:45pm

Authors:
Leproust, E., Twist Bioscience
Twist Bioscience is revolutionizing DNA Synthesis. Our proprietary semiconductor-based technology enables rapid and highly parallel synthesis of hundreds of thousands of the highest quality oligonucleotides. The oligos can be pooled for genome editing applications or assembled to generate your favorite genes or libraries.

A Butyrate-Producing Synthetic Biotic to Treat Inflammatory Bowel Disease In Vivo

- Conference: Synthetic Biology Engineering Evolution Design SEED
- Year: 2017
- Proceeding: 2017 Synthetic Biology: Engineering, Evolution & Design (SEED)
- Group: General Submissions
- Session:

  Session 4: Microbiome Engineering

- Time:

  Tuesday, June 20, 2017 - 4:15pm-4:45pm

Authors:
Fisher, A., Synlogic Therapeutics
Isabella, V. M., Synlogic Inc.
Miller, P., Synlogic Inc.
Recent studies into the interplay between the microbiome and host have extensively implicated short-chain fatty acid (SCFA) fermentation products as contributors to both intra and extra intestinal health. In the context of the gut, butyrate (a four carbon SCFA) serves as the primary energy source for epithelial colonocytes, drives regulatory T cell differentiation and is therefore essential for the maintenance of colonic mucosal health. On a systemic level, butyrate may influence insulin sensitivity, cholesterol synthesis and autoimmunity. Typically, butyrate is produced by the action of slow-growing Gram positive organisms â□“ often Clostridial in origin â□“ during the fermentation of cellulose and complex carbohydrates. To counteract the depletion of butyrate in the gut, we iteratively engineered *Escherichia coli* Nissle 1917 to produce butyrate. The new butyrate-producing Nissle was capable of millimolar level production of butyrate comparable with the levels produced by enteric Clostridial species. The circuitry was placed under anaerobic control to support induction upon entry into the gastrointestinal track. Finally, proof-of-concept *in vivo* experimentation showed production of butyrate in the mouse gut. This engineered strain may be useful for the treatment of a variety of gastrointestinal and systemic conditions associated with SCFA deficiency.

Engineering the Mammalian Gut Microbiome in Situ

- Conference: Synthetic Biology Engineering Evolution Design SEED
- Year: 2017
- Proceeding: 2017 Synthetic Biology: Engineering, Evolution & Design (SEED)
- Group: General Submissions
- Session:

  Session 4: Microbiome Engineering

- Time:

  Tuesday, June 20, 2017 - 4:45pm-5:15pm

Authors:
Wang, H. H., Columbia University
Ronda, C., Columbia University
Chen, S., Columbia University
Cabral, V., Columbia University
Engineering microbial populations in open environments, such as the mammalian gastrointestinal tract, is an outstanding challenge. Current approaches require laboratory cultivation and tedious optimizations to develop genetic tools for each new non-model microbe. Here, we describe an enabling platform, Gene Transfer In Situ (GeTIS), to identify and engineer genetically tractable yet undomesticated microbes from complex microbial communities. Using engineered horizontal gene transfer and transposon systems, we identified and isolated bacteria from the mammalian gut microbiome that were amenable to genetic manipulation and then redeployed them back into the mammalian host as host-optimized engineerable probiotics. Furthermore, we demonstrated *in situ* gene transfer in a live animal to deliver new traits directly into its established gut microbiota. This *in situ* genome engineering approach enables the accelerated development of new microbial chasses for synthetic biology and the introduction of novel capabilities into established microbial communities with minimal disruption to their native milieu.

Paper-Based Diagnostic for Assessing the Gut Microbiome

- Conference: Synthetic Biology Engineering Evolution Design SEED
- Year: 2017
- Proceeding: 2017 Synthetic Biology: Engineering, Evolution & Design (SEED)
- Group: General Submissions
- Session:

  Session 4: Microbiome Engineering

- Time:

  Tuesday, June 20, 2017 - 5:15pm-5:45pm

Authors:
Takahashi, M. K., Cornell University
Tan, X.
Dy, A. J., MIT
Braff, D.
Collins, J., Boston University
New studies are showing connections between the relative abundances of the bacterial species in the human gut to many health conditions including inflammatory bowel disease, obesity, autism, and cancer. In most cases, the precise mechanism by which alterations in the microbiome affect these various medical conditions is still being investigated. Current methods for profiling the microbiome typically involve deep sequencing of processed stool samples coupled with high throughput bioinformatics analysis of the sequences. These techniques are expensive, slow, and require significant technical expertise to design, run, and interpret. These limitations have severely restricted the large-scale prospective monitoring of patient cohorts necessary to determine the causal relationship between the microbiome and human health. Here we present a paper-based molecular diagnostic platform for simple and affordable analysis of the gut microbiome. The core of the platform is two technologies â□" RNA toehold switch sensors and an *in vitro* cell-free transcription-translation system freeze-dried onto small paper discs. When combined, these two technologies provide the ability to detect nearly any RNA sequence with a simple fluorescence or colorimetric output, for approximately $1 per reaction. Here we design toehold switch sensors to detect the V3 hypervariable region of the 16S ribosomal RNA (rRNA) of a panel of microbes found in the human gut. We demonstrate good specificity of the 16S rRNA sensors and show that specificity can be improved by targeting species specific mRNA transcripts found through bioinformatic analysis. In order to achieve the necessary sensitivity to analyze human fecal samples we employ NASBA (nucleic acid sequenced based amplification) to isothermally amplify

specific RNAs prior to detection by our toehold switch sensors. Finally, we show that the toehold switch sensors can be used to quantify amplification by NASBA and validate our measurements of bacterial RNA from total RNA extracted from human fecal samples using RT-qPCR.

Microbial Delivery of Antimicrobial Peptides for Protection Against Vancomycin-Resistant Enterococcus

- Conference: Synthetic Biology Engineering Evolution Design SEED
- Year: 2017
- Proceeding: 2017 Synthetic Biology: Engineering, Evolution & Design (SEED)
- Group: General Submissions
- Session:

  Session 4: Microbiome Engineering

- Time:

  Tuesday, June 20, 2017 - 5:45pm-6:00pm

Authors:
Geldart, K., University of Minnesota
Forkus, B., University of Minnesota
Kaznessis, Y. N., University of Minnesota
Antimicrobial peptides (AMPs) are naturally produced by our microflora as a first line of defense against invading species. While numerous probiotic bacterial species are known to produce AMPs targeting different pathogens of interest, native peptide expression is often low or induced under uncontrollable conditions making these probiotics therapeutically unreliable. We harness this natural defense mechanism for the treatment of antibiotic resistant intestinal infections by engineering probiotic bacteria that can be orally administered and will reliably express and secrete AMPs targeting the pathogen of interest in the intestines. We are developing libraries of probiotic species, promoters, secretion systems, and antimicrobial peptides to enable rapid development of new probiotics targeting a wide array of pathogens.[1,2,3,4] Herein, we engineer probiotic *E. coli* Nissle 1917 (EcN) to deliver AMPs targeting vancomycin-resistant *Enterococcus* (VRE), an opportunistic pathogen that causes thousands of hospital-acquired infections and deaths each year.

We have developed an AMP expression vector for Nissle that uses a strong synthetic promoter to drive the expression of three previously characterized anti-enterococcal AMPs, Enterocin A, Enterocin B, and Hiracin JM79.[5] This vector contains the secretion machinery from Microcin V, an *E. coli*-derived AMP, to enable secretion of the anti-enterococcal peptides.[4] We first demonstrate the ability of the probiotic to reduce VR *E. faecium* and *E. faecalis* counts in laboratory cultures using both solid and liquid culture activity assays. EcN expressing anti-enterococcal AMPs eliminates over 99% of VR *E. faecium*after one hour of co-culture.

We then test the efficacy of our probiotics in reducing intestinal pathogen counts in mice colonized with VRE via oral administration. Mice administered EcN producing AMPs on average exhibited a 10x reduction in *E. faecium* fecal counts compared to mice administered no probiotic. This reduction of VRE in the treated versus untreated groups was significant with $p = 0.011$ based on a Studentâ□™s t-test comparing total *E. faecium* fecal levels over time. This work provides proof-of concept evidence supporting the potential use of AMP-producing probiotic bacteria for the reduction of VRE in the intestines.

Importantly, we demonstrate that this EcN probiotic platform can be readily modified to express alternative peptides targeting a variety of pathogens including *Salmonella* and diarrheagenic *E. coli*. Additionally, unlike broad-spectrum traditional antibiotics, the AMPs expressed by these microbes offer species-specific activity. This specificity reduces off-target pressure for resistance development and supports competitive exclusion of the pathogen by the surrounding microbiota.

1. Volzing, ACS Synth Biol. 2013; 2(11):643-50

2. Geldart, Appl Environ Microbiol. 2015; 81(11): 3889â□"3897.

3. Forkus, Sci Rep. 2017; 7: 40695.

4. Geldart, Pharmaceuticals (Basel). 2016; 9(4): 60.

5. Geldart, Antimicrob Agents Chemother. 2017; AAC.02033-16

Engineering Next-Generation T Cells for Cancer Immunotherapy

- Conference: Synthetic Biology Engineering Evolution Design SEED
- Year: 2017
- Proceeding: 2017 Synthetic Biology: Engineering, Evolution & Design (SEED)
- Group: General Submissions
- Session:

  Keynote 2

- Time:

  Wednesday, June 21, 2017 - 8:30am-9:30am

Authors:
Chen, Y. Y., University of California, Los Angeles
The adoptive transfer of T cells expressing chimeric antigen receptors (CARs) has demonstrated clinical efficacy in the treatment of advanced cancers, with anti-CD19 CAR-T cells achieving up to 90% complete remission among patients with relapsed B-cell malignancies. However, challenges such as antigen escape and immunosuppression limit the long-term efficacy of adoptive T-cell therapy. Here, I will discuss the development of next-generation T cells that can target multiple cancer antigens and resist immunosuppression, thereby increasing the robustness of therapeutic T cells against tumor defense mechanisms. This presentation will highlight the potential of synthetic biology in generating novel mammalian cell systems with multifunctional outputs for therapeutic applications.

Pipeline Technology from Oxitec Ltd: Genetic Control of Mosquito Vectors and Crop Pests

- Conference: Synthetic Biology Engineering Evolution Design SEED
- Year: 2017
- Proceeding: 2017 Synthetic Biology: Engineering, Evolution & Dcsign (SEED)
- Group: General Submissions
- Session:

  Session 5: Applications of Synthetic Biology

- Time:

  Wednesday, June 21, 2017 - 10:30am-11:00am

Authors:
Rose, N.

Mosquito-borne diseases such as Zika, dengue fever and Chikungunya are major and increasing international public health concerns. The primary vector of these viral diseases is *Aedes aegypti* and current control measures against this mosquito are proving inadequate or problematic. An innovative and effective control technique for *Aedes aegypti* involves the release of self-limiting mosquitoes. These mosquitoes are genetically engineered to be homozygous for a self-limiting gene under control of an on/off tetracycline switch. When released into the wild, any offspring of the transgenic males will inherit one copy of the transgene. In the absence of tetracycline antidote, the transgene will prevent the mosquito from developing to adulthood, thus suppressing the population with repeated releases. Oxitecâ□™s self-limiting transgenic insect technology has also been applied successfully to develop population control measures for major crop pests such as the Mediterranean fruit fly (Medfly) and the diamondback moth. In a variation on the self-limiting genetics used in *Aedes* mosquitoes, genetic sexing strains that express the self-limiting gene in a sex-specific manner are used to produce male-only cohorts of these crop pests for release in control programmes. These and other developments in Oxitecâ□™s genetic engineering programme to enable biological control of major insect pests will be discussed. The Oxitec transgenic mosquito method has dramatically reduced *Aedes aegypti* populations in trials around the world and is now being used under operational conditions in the Cayman Islands and Brazil to control *Aedes aegypti*. Results of these programmes will be presented.

Towards Safe and Compatible Organs for Transplantation

- Conference: Synthetic Biology Engineering Evolution Design SEED
- Year: 2017
- Proceeding: 2017 Synthetic Biology: Engineering, Evolution & Design (SEED)
- Group: General Submissions
- Session:

  Session 5: Applications of Synthetic Biology

- Time:

  Wednesday, June 21, 2017 - 11:00am-11:30am

Authors:
Güell, M., Wyss Institute for Biologically Inspired Engineering
The shortage of organs for transplantation is a major barrier to the treatment of organ failure. Although porcine organs are considered promising resources for human transplant, their use has been checked by concerns about the transmission of porcine endogenous retroviruses (PERVs) to humans and pig-human immunological incompatibility. We engineered primary cell lines at multiple loci, and used somatic cell nuclear transfer to produce pigs with advanced genotypes for xenotransplantation.

Orthogonal Pair Directed, Codon Specific, Sense Codon Reassignment: An Improved Tool for Evaluating the Plasticity of the E coli Genetic Code

- Conference: Synthetic Biology Engineering Evolution Design SEED
- Year: 2017
- Proceeding: 2017 Synthetic Biology: Engineering, Evolution & Design (SEED)
- Group: General Submissions
- Session:

  Session 6: Re-writing the Genetic Code

- Time:

  Wednesday, June 21, 2017 - 2:15pm-2:45pm

Authors:
Fisk, J. D., Colorado State University

Natural proteins are constructed out of 20 amino acids building blocks which contain a limited set of chemical functionalities. Expanding the set of genetically encoded amino acids introduces new chemical moieties that enable the specific modification, labeling, and cross-linking of proteins. Genetically encoding new amino acids with diverse chemical functionalities is limited by the fact that all 64 triplet codons have assigned functions in the genetic code. Two general methods of genetic code expansion are commonly employed, but each has limitations. Non-canonical amino acid residues can be incorporated in response to stop codons, but this strategy is limited to single site incorporation. Residue specific reassignment allows multisite incorporation, but requires complete substitution of one natural amino acid for a non-canonical residue. A third approach combining aspects of both non-sense and residue specific reassignment enables the expansion of the genetic code by reassigning the meaning of sense codons.

We have developed a fluorescence-based screen to quantify the extent to which the sense codons in E coli can be reassigned by introducing *M. jannaschii* tyrosine and *M. barkeri* pyrrolysine orthogonal aminoacyl tRNA synthetase/tRNA pairs directed to decode sense codons. The *M. jannaschii* tyrosine and *M. barkeri* pyrrolysine orthogonal pairs are commonly used to incorporate non-cannonical amino acids in response to nonsense codons. The fluorescence screen was employed to examine the complete set of codons read via wobble interactions as well as the set of rarely used codons in E coli. The method of using orthogonal pairs to precisely direct the incorporation of amino acids in response to sense codons allows a more focused examination of the physiological effects of substitutions than previous methods employing aminoacyl tRNA synthetases with defective editing sites or ribosomes with relaxed proofreading. We describe the efficiency and specificity of sense codon reassignment of *E. coli* codons and the ability to improve these qualities through the application of the fluorescence-based screen in a high throughput mode to select improved variants from large libraries. The main finding of these studies are that the E coli genetic code is readily reassignable and genome wide amino acid substitutions are well tolerated. We observe weak correlations between tRNA abundance and changes in amino acylation efficiency of the modified orthogonal system on sense codon reassignment. The relative energetics of codon -- anticodon interactions appear to be an important factor in reassignment efficiency. We identify four codons that have not been targeted for sense codon reassignment that appear to promising.

Controlling the Metabolic Labeling of Proteins with Non-Natural Amino Acids Using Engineered Ligand-Responsive Amino Acyl tRNA Synthetases

- Conference: Synthetic Biology Engineering Evolution Design SEED
- Year: 2017
- Proceeding: 2017 Synthetic Biology: Engineering, Evolution & Design (SEED)
- Group: General Submissions
- Session:

  Session 6: Re-writing the Genetic Code

- Time:

  Wednesday, June 21, 2017 - 2:45pm-3:00pm

Authors:
Thomas, E.
Silberg, J., Rice University
A wide variety of amino acyl tRNA synthetases (aaRSs) have had their substrate specificities altered through protein engineering to create mutant aaRSs that can charge tRNA with non-canonical amino acids in cells. While the activities of these mutant aaRSs can be dynamically controlled in cells using conditional promoters, aaRS can also be synthesized using non-natural amino acids as they accumulate in cells, which in some cases could negatively impact their ability to metabolically label other proteins. In addition, our ability to quickly switch aaRS-mediated metabolic labeling on and off remains limited because we have not yet discovered ways to directly control aaRS activity post-translationally in response to environmental changes. To develop prototype aaRSs with activities that are dependent upon direct binding to a chemical, we have engineered a first generation of ligand-responsive aaRSs. For these efforts, we targeted a mutant *Escherichia coli* methionyl tRNA synthetase (MetRS), which was previously developed to charge tRNA with azidonorleucine (Anl) in *Escherichia coli*. We first used a combinatorial approach to discover split MetRS whose fragments cooperatively function when each fragment is fused to a pair of proteins that interact. We then examined whether the activity of split MetRS can be regulated by either fusing the fragments to a pair of proteins whose interaction is stabilized by ligand binding or by fusing the fragments to the termini of a single protein domain that exhibits a ligand-dependent conformational change. We have found that both approaches can be used to regulate Anl labeling of newly synthesized proteins in a ligand-dependent manner. When

MetRS fragments were fused to FKBP12 and the rapamycin binding domain of mTOR, metabolic labeling was significantly enhanced in growth medium containing rapamycin, which stabilizes the FKBP12-mTOR complex. In addition, fusion of MetRS fragments to the termini of the ligand-binding domain of the estrogen receptor yielded a protein whose metabolic labeling was significantly enhanced when 4-hydroxytamoxifen was added to the growth medium. Ongoing efforts are focused on assessing whether the ligand-dependent metabolic labeling mediated by our engineered proteins can be switched on faster than labeling achieved using transcriptional regulation, improving the metabolic labeling efficiencies that can be achieved with our prototype aaRS switches, and investigating the diversity of ligands that can be used to directly regulate MetRS activity. The protein switches arising from these studies are expected to be useful for extending our control over protein metabolic labeling with Anl by allowing for fast regulation of this labeling using post-translational reactions. Furthermore, this approach can be applied to metabolic labeling with additional non-natural amino acids by extending our designs to structurally-related aaRS with distinct substrate specificities. Ligand-responsive aaRSs are expected to have applications in the production of protein biomaterials, proteomic studies, and protein pharmaceutical biosynthesis.

•

Engineering and Optimizing Biofuel-Tolerant Microbes

- Conference: Synthetic Biology Engineering Evolution Design SEED
- Year: 2017
- Proceeding: 2017 Synthetic Biology: Engineering, Evolution & Design (SEED)
- Group: General Submissions
- Session:

  ACS Synthetic Biology Young Investigator Award Lecture

- Time:

  Wednesday, June 21, 2017 - 3:00pm-3:30pm

Authors:
Dunlop, M. J., N/A
Microbes can be engineered to produce biologically-derived replacements for gasoline, diesel, and jet fuel. However, a major technical challenge is that cells can only tolerate limited concentrations of typical fuels. Hydrocarbons are toxic because they damage the cell membrane and increase membrane fluidity. Thus, without further engineering there is a fundamental upper limit to how much biofuel microbes can produce, and cells must trade off production against survival. However, there are many natural examples of microbes that thrive in oil rich environments, such as around natural oil seeps or near oil spills. These microbes harbor mechanisms for tolerating the toxic effects of hydrocarbons. By using library-based approaches, we have identified several novel strategies for improving microbial biofuel tolerance by selecting for genes that enable survival. For example, expression of efflux pumps can dramatically improve tolerance in production strains, and we have further improved survival by designing synthetic feedback systems to control pump expression. This talk will discuss our efforts to enhance microbial synthesis of next-generation biofuels by developing tools for improving microbial tolerance of biofuel production conditions.

Universal Chimeric Antigen Receptors for Multiplexed and Logic Control of T Cell Responses

- Conference: Synthetic Biology Engineering Evolution Design SEED
- Year: 2017
- Proceeding: 2017 Synthetic Biology: Engineering, Evolution & Design (SEED)
- Group: General Submissions
- Session:

  Session 7: Mammalian Synthetic Biology

- Time:

  Wednesday, June 21, 2017 - 4:30pm-5:00pm

Authors:
Wong, W., Boston University

T cells expressing chimeric antigen receptors (CAR) have shown promises in clinical trials against various cancers, but can also have toxicities due to off-targeting and over-activation. To improve the specificity, flexibility, and precision of the CAR, we developed a split, universal, and programmable (SUPRA) CAR system composed of a universal receptor expressed on T cells and a tumor-targeting scFv adaptor molecule. The universal receptor is the fusion of intracellular signaling domains and a leucine zipper as the extracellular domain (zipCAR). The adaptor molecule is the fusion of a cognate leucine zipper and a scFv (zipFv). The scFv on the zipFv binds to the antigen and the leucine zipper binds and activates the zipCAR on the T cells. We showed that our system has an ON/OFF switch functionality, can flexibility tune T cell activation, and can sense multiple antigens to perform logic computations. By developing orthogonal SUPRA CAR systems, we showed that we can independently regulate multiple signaling pathways in the same cell, or separately control immune responses from different cell types. We envision that this work will provide much-needed tools to improve the safety and efficacy of cellular cancer immunotherapy.

Programming Protein Function with Synthetic RNA Splicing Devices

- Conference: Synthetic Biology Engineering Evolution Design SEED
- Year: 2017
- Proceeding: 2017 Synthetic Biology: Engineering, Evolution & Design (SEED)
- Group: General Submissions
- Session:

  Session 7: Mammalian Synthetic Biology

- Time:

  Wednesday, June 21, 2017 - 5:00pm-5:15pm

Authors:
Mathur, M., Stanford University
Kim, C., Stanford University
Munro, S., National Institute of Standards and Technology
Smolke, C. D., Stanford University
Nature offers enormous functional diversity that is largely untapped in engineered biological systems. Given its roots in bacterial transcriptional gene networks, synthetic biology has yet to develop a comprehensive set of tools that harness the complexity and biological regulation prevalent in mammalian cells. A critically important mechanism in mammalian cells, alternative RNA splicing, increases the coding capacity of genomes to generate several proteins from an individual gene. Few engineered genetic devices utilizing alternative splicing mechanisms have been described, and existing approaches are limited to turning on or off the expression of a gene coupled to a splicing regulatory component. A novel alternative splicing platform has the potential to broaden the current regulatory capabilities of genetic devices by providing a strategy to program protein function. We describe the design and implementation of an intron framework for controlling the mutually exclusive alternative splicing of exons in a four exon-three intron system, wherein the exon sequences can be recoded. The coding capacity of the genetic device is increased by the mutually exclusive nature of the alternative splicing event, as one of two internal exons is incorporated into a mature mRNA, thereby resulting in multiple isoforms. RNA profiling and protein expression assays established that these devices effectively spliced to produce the expected dominant isoform. Systematic tuning of consensus sequence element activity in the intron framework revealed the ability to modify isoform profiles and favor alternate isoform production by a device. Furthermore, we conducted an exploratory study of intronic regulatory elements for controlling and predicting mutually exclusive alternative splicing events within the context of the intron framework. Finally, we applied this intron framework to program

transcription factor activity by encoding domains of modular transcription factors in synthetic RNA splicing devices, modifying domain functionality through alternative splicing, and producing precise transcription factor profiles that control gene expression patterns. Our work highlights the first demonstration of an intron framework design that leverages a complex mode of alternative splicing to program protein function with synthetic RNA splicing devices. Continued development of such foundational frameworks in mammalian synthetic biology will enable increasingly sophisticated devices that expand the capabilities of genetic engineering for tackling critical global challenges in health and medicine.

Understanding and Mitigating the Effects of Limited Transcriptional Resources in Mammalian Cells

- Conference: Synthetic Biology Engineering Evolution Design SEED
- Year: 2017
- Proceeding: 2017 Synthetic Biology: Engineering, Evolution & Design (SEED)
- Group: General Submissions
- Session:

  Session 7: Mammalian Synthetic Biology

- Time:

  Wednesday, June 21, 2017 - 5:15pm-5:30pm

Authors:
Jones, R., MIT
Qian, Y., MIT
Weiss, R., M.I.T.
Del Vecchio, D., MIT

Mammalian synthetic biologists seek to design molecular circuits with many potential therapeutic and biotechnology applications. Developing these often complex devices requires being able to reliably compose multiple modular regulatory modules in varying cellular contexts. However, predicting module behavior can be difficult due to â□œhiddenâ□□ or unpredictable interactions between circuit parts. Parts competing for limiting cellular factors can be a significant contributor to these hidden interactions. In eukaryotes, *trans*-activation domains of transcription factors can bind and sequester co-activator proteins, causing effective repression of otherwise active promoters in a phenomenon known as â□œsquelchingâ□□ [1, 2]. Since many synthetic mammalian circuits are based on transcriptional networks, understanding and mitigating the effects of squelching will allow for the design of more reliable and predictable circuitry for many applications.

To understand squelching, we built an extended model of transcription, which includes a pool of conserved general co-activator resources required for transcription initiation at each promoter. We then experimentally characterized the degree of competition between commonly used promoter and activator parts to parameterize the effective repression due to resource competition in the model. The experimental results show that promoters and activators can both compete with each other and among themselves to a significant degree. Further, we have shown the dose-dependence of titrating in competitors for select combinations of promoters and activators.

To mitigate the effects of squelching, we designed a novel phosphorylation-based feedback module, which aims to buffer against resource competition within reasonable competitor input regimes. The circuit is composed of a kinase input that phosphorylates a transcription factor, allowing it to bind DNA and initiate transcription of a target promoter. A phosphatase is co-expressed with the output, and de-phosphorylates the transcription factor. The feedback works on the principle of pseudo-integral feedback, which theoretically allows the target protein concentration to be independent of resource availability. Simulation results show that this feedback module works for buffering out resource competition effects, and we are in the process of experimental characterization. The module parts are implemented with *E. coli* two-component signaling histidine kinases and phosphatases, which have been shown to work in mammalian cells when the response regulator is fused to a eukaryotic activation domain such as VP16 [3]. We have applied conserved mutations [4] and domain rearrangements [5, 6] to convert bifunctional histidine kinases to variants with only kinase or phosphatase activity, rather than both. We have verified that these variants work as predicted in HEK 293FT cells, and have achieved up to 400-fold changes in promoter output between when a phosphatase or kinase variant are present. We are currently pursuing experiments to determine under what regimes the feedback is effective in buffering against resource competition.

[1] Gill G & Ptashne M. 1988. *Nature*. 334: 721-724,

[2] Sadowski I *et al.* 1990. *Nature*. 335: 563-564.

[3] Hansen J *et al.* 2014. *PNAS*. 111: 15705-15710.

[4] Willett J & Kirby J. 2012. *PLoS Genetics*. 8: e1003084.

[5] Zhu Y *et al.* 2000. *PNAS*. 97: 7808-7813.

[6] Qin L *et al.* 2000. *Mol Microbiol*. 36: 24-32.

Harnessing the Chemistry of Plant Natural Product Biosynthesis

- Conference: Synthetic Biology Engineering Evolution Design SEED
- Year: 2017
- Proceeding: 2017 Synthetic Biology: Engineering, Evolution & Design (SEED)
- Group: General Submissions
- Session:

  Keynote 3

- Time:

  Thursday, June 22, 2017 - 8:30am-9:30am

Authors:
O'Connor, S.

Plants, which make thousands of complex natural products, are outstanding chemists. Through the concerted action of enzymes that are assembled into metabolic pathways, nature creates chemical complexity from simple starting materials. I will highlight some of the unusual enzymatic transformations that plants use to make complex, bioactive natural products, and will also discuss methods by which these pathways can be harnessed for metabolic engineering. The focus is on the biosynthesis of the monoterpenes called iridoids, and the alkaloid derived from iridoids, known as the monoterpene indole alkaloids. The discovery, functional characterization and mechanistic study of enzymes involved in the biosynthesis of these important compounds in several medicinal plant species will be discussed.

Conversion of Glycerol to 3-Hydroxypropanoic Acid By Genetically Engineered Bacillus Subtilis

- Conference: Synthetic Biology Engineering Evolution Design SEED
- Year: 2017
- Proceeding: 2017 Synthetic Biology: Engineering, Evolution & Design (SEED)
- Group: General Submissions
- Session:

  Session 8: Metabolic Engineering

- Time:

  Thursday, June 22, 2017 - 10:00am-10:30am

Authors:
Kalantari, A., Chalmers University of Technology
Ji, B., Chalmers University of Technology
Chen, T.
Mijakovic, I., Chalmers University of Technology

3-hydroxypropanoic acid (3-HP) is an important biomass-derivable platform chemical that can be converted into a number of industrially relevant compounds. There have been several attempts to produce 3-HP from renewable sources in cell factories, focusing mainly on *Escherichia coli*, *Klebsiella pneumoniae* and *Saccharomyces cerevisiae*. Despite the significant progress made in this field, commercially exploitable large-scale production of 3-HP in microbial strains has still not been achieved. In this study we investigated the potential of *Bacillus subtilis* as a microbial platform for bioconversion of glycerol into 3-HP. Our recombinant *B. subtilis* strains overexpress the two-step heterologous pathway containing glycerol dehydratase and aldehyde dehydrogenase from *K. pneumoniae*. Genetic engineering, driven by *in silico* optimization, and optimization of cultivation conditions resulted in a 3-HP titer of 10 g/L, in a standard batch cultivation. Our findings provide the first report of successful introduction of the biosynthetic pathway for conversion of glycerol into 3-HP in *B. subtilis*. With this relatively high titer in batch, and the robustness of *B. subtilis* in high density fermentation conditions, we expect that our production strains may constitute a solid basis for commercial production of 3-HP.

Improving E. coli Methyl Ketone Production during Fed-Batch Fermentation By Co-Factor Balancing

- Conference: Synthetic Biology Engineering Evolution Design SEED
- Year: 2017
- Proceeding: 2017 Synthetic Biology: Engineering, Evolution & Design (SEED)
- Group: General Submissions
- Session:

  Session 8: Metabolic Engineering

- Time:

  Thursday, June 22, 2017 - 10:30am-11:00am

Authors:
Goh, E. B., Joint BioEnergy Institute (JBEI)
Baidoo, E. E. K., Joint BioEnergy Institute
Ando, D. E., Joint BioEnergy Institute
Keasling, J. D., Joint BioEnergy Institute (JBEI)
Garcia Martin, H., Joint BioEnergy Institute (JBEI)
Beller, H. R., Joint BioEnergy Institute (JBEI)

We have engineered *Escherichia coli* to overproduce aliphatic methyl ketones (MK) in the C_{11} to C_{15} (diesel) range; this group of MK includes 2-undecanone and 2-tridecanone, which have favorable cetane numbers and are also of importance to the flavor and fragrance industry. Overall, we have made specific improvements that resulted in 50,000-fold enhancement in MK titer relative to that of a fatty acid-overproducing *E. coli* strain. The first generation of engineered *E. coli* (Goh et al. 2012) had modifications that included: (a) overproduction of Î²-ketoacyl-coenzyme A (CoA) thioesters achieved by modification of the Î²-oxidation pathway (specifically, overexpression of a heterologous acyl-CoA oxidase and native FadB, and chromosomal deletion of *fadA*) and (b) overexpression of a native thioesterase (FadM). Subsequent genetic modifications included balancing overexpression of *fadR* and *fadD* to increase fatty acid flux into the pathway, consolidation of the pathway from two plasmids into one, codon optimization, and knocking out key acetate production pathways (resulting in strain EGS1895). This strain produced an MK titer of 1.4 g/L with 1% glucose in shake-flask experiments, representing 40% of the maximum theoretical yield. ^{13}C-based metabolic flux analysis with EGS1895 showed increased carbon flux through the pentose phosphate pathway (PPP), which we hypothesized is to generate more NADPH - an important co-factor for FabG (Î²-ketoacyl-ACP reductase) activity during fatty acid biosynthesis.Â In an attempt to divert flux away from PPP, we overexpressed alternate FabG variants that we have demonstrated preferentially use NADH instead of NADPH. Specifically, overexpression of the NADH-dependent *Acholeplasma laidlawii* FabG increased MK production by 20-50% in shake-flask experiments and had more sustained MK production in fed-batch fermentation compared to EGS1895. We have achieved MK titers of ~6.0 g/L in fed-batch fermentation, which is the best MK titer observed to date.

Production of Industrial Products from Natural Gas Using an Engineered Methanotroph Biocatalyst

- Conference: Synthetic Biology Engineering Evolution Design SEED
- Year: 2017
- Proceeding: 2017 Synthetic Biology: Engineering, Evolution & Design (SEED)
- Group: General Submissions
- Session:

  Session 8: Metabolic Engineering

- Time:

  Thursday, June 22, 2017 - 11:00am-11:30am

Authors:
Held, M.
Kealey, J.
Natural gas is one of the most economical and abundant sources of carbon currently available, and is therefore an attractive feedstock for the production of industrial products by fermentation. Methanotrophic bacteria use methane in natural gas as a source of carbon and energy and have been grown at industrial scale for production of biomass for animal feed. Due to a paucity of genetic tools available for methanotroph systems, methanotroph bacteria have not been extensively engineered for the production of bio-based heterologous products. We have developed a powerful suite of technologies that enables the rapid engineering and screening of methanotroph biocatalysts and applied it for the production of a variety of heterologous products from natural gas. We will describe challenges in developing this platform and the progress of engineering non-model methanotroph hosts for production of industrial products by gas fermentation.

Directed Evolution of Targeted Cas9 to Specific Cleavage Sites

- Conference: Synthetic Biology Engineering Evolution Design SEED
- Year: 2017
- Proceeding: 2017 Synthetic Biology: Engineering, Evolution & Design (SEED)
- Group: General Submissions
- Session:

    Session 9: Genome Editing

- Time:

    Thursday, June 22, 2017 - 1:30pm-2:00pm

Authors:
Steinberg, B., Editas Medicine
Cerchione, D.
Maeder, M.
Skor, M.
Jayaram, H.
Myer, V.
CRISPR/Cas9 has wide-reaching scientific and therapeutic applications, but efficient, exact targeting of cutting sites can be challenging. Several cas9 variants have been engineered to target alternate PAMs; however, generation of new variants is often laborious. Additionally, the cutting ability of cas9 nucleases is often highly guide-dependent and shows a large amount of variation between sites. In order to engineer new nuclease variants targeted to cleave a given site, we developed a directed evolution platform to select for DNA cleavage in bacteria. Our system allows rapid generation of mutant libraries, which are challenged with phage containing target sites in a competitive pool. Using this method, we developed *S. aureus* cas9 variants which have altered PAM preferences to target cleavage to novel targets of therapeutic relevance.

CRISPR Technologies 2.0: Harnessing the Next Generation of Cas Nucleases for Genome Editing and Beyond

- Conference: Synthetic Biology Engineering Evolution Design SEED
- Year: 2017
- Proceeding: 2017 Synthetic Biology: Engineering, Evolution & Design (SEED)
- Group: General Submissions
- Session:

  Session 9: Genome Editing

- Time:

  Thursday, June 22, 2017 - 2:00pm-2:30pm

Authors:
Beisel, C. L., North Carolina State University
CRISPR technologies represent revolutionary tools for gene editing in basic research, biotechnology, medicine, and agriculture and have driven ethical debates around GMOâ□™s and editing human germline cells. Despite the fanfare surrounding these technologies, they have a common, natural source: adaptive immune systems in bacteria and archaea called CRISPR-Cas systems. CRISPR-Cas systems use guide RNAs to direct Cas nucleases to bind and cleave complementary genetic material. Guide RNAs can be readily designed de novo to direct the binding and cleavage of specific DNA sequences, lending to gene editing and many other applications. While existing CRISPR technologies have overwhelmingly relied on the Cas9 nuclease, nature offers a myriad of other Cas proteins with unique mechanisms and functions. Harnessing these more recently discovered nucleases holds the potential to improve and expand existing CRISPR technologies and bring about a further biotechnological revolution.

Here, I will discuss my groupâ□™s work on the recently discovered Type V-A Cas nuclease Cpf1. Like the standard *S. pyogenes* Cas9, Cpf1 is a single effector protein that is directed by guide RNAs to cleave target DNA. However, unlike Cas9, Cpf1 is smaller, generates an overhang cleavage product, and can process a native CRISPR array into individual guide RNAs without any accessory factors. These features could allow Cpf1 to eventually displace Cas9 across its many uses. We have been rapidly prototyping Cpf1 and other novel Cas nucleases using cell-free TXTL systems, generating design rules for genome editing and programmable gene regulation. We have also developed an efficient, one-pot cloning scheme for large CRISPR arrays for multiplexing applications. These advances are transitioning Cpf1 and other recently discovered Cas nucleases into a new generation of CRISPR technologies for emerging applications in genome editing, scalable gene circuit design, and high-throughput phenotypic screens.

Rational Engineering of Chimaeric Recombinases for Genomic Excision

- Conference: Synthetic Biology Engineering Evolution Design SEED
- Year: 2017
- Proceeding: 2017 Synthetic Biology: Engineering, Evolution & Design (SEED)
- Group: General Submissions
- Session:

  Session 9: Genome Editing

- Time:

  Thursday, June 22, 2017 - 2:30pm-2:45pm

Authors:
Abioye, J. A., University of Glasgow
Stark, W. M., University of Glasgow
Olorunniji, F. J., University of Glasgow
Reboud, J., University of Glasgow
Cooper, J. M., University of Glasgow
Holt, S. E., University of Glasgow
Proudfoot, C. M., University of Glasgow
Most current genome editing approaches being developed as potential therapeutic strategies utilize nuclease-based systems such as the TAL effector nucleases, Zinc-finger nucleases and CRISPR-Cas9. The mechanisms of action of these involve cleavage followed by homology-directed repair (HDR) or non-homologous end joining (NHEJ) at the cleavage site. However, off-target activities and fatal indel mutations characteristic of these approaches could have debilitating effects on the host cells and reduce the effectiveness of their application for precise gene therapy. Here, we present the rational engineering and in vitro characterization of Tal-effector recombinases (TALERs) capable of specific excision and religation on non-cognate sequences. These chimaeric recombinases comprise an N-terminal serine recombinase catalytic domain linked to a TAL effector C-terminal DNA-binding domain. We describe the essential TALER architecture, target site requirements and their in vitro recombination properties. As a proof of concept, we have engineered components of chimaeric recombinases to recognize and promote recombination at a highly conserved region within the HIV-1 long terminal repeats (LTR) flanking the proviral DNA at both ends. Each target sequence (one at each end of the provirus DNA) will be bound by two recombinase subunits, and then the two sites will be brought together, cleaved and recombined. As a result, the provirus is excised from the host genome (leaving behind a non-functional â□œscarâ□□ sequence). The requirement for interaction of two recombinase-bound sites, and the lack

of necessity for any host cell-encoded factors should maximize the fidelity and efficiency of provirus removal. Our research to date has used a combination of random mutagenesis and rational engineering approaches to create a Tn3 resolvase-based recombinase catalytic domain capable of targeting the central part of our chosen HIV target sequence. We have also deduced the optimal structure and design of functional TAL effector chimaeric recombinases. Our current work is focused on assembling and characterizing the complete HIV-TAL effector recombinase and the validation of the engineered synthetic protein for in vivo genome editing.

Twin-Primer Non-Enzymatic DNA Assembly: An Efficient and Accurate Multi-Part DNA Assembly Method

- Conference: Synthetic Biology Engineering Evolution Design SEED
- Year: 2017
- Proceeding: 2017 Synthetic Biology: Engineering, Evolution & Design (SEED)
- Group: General Submissions
- Session:

  Session 10: Genetic Engineering Tools

- Time:

  Thursday, June 22, 2017 - 3:45pm-4:15pm

Authors:
Liang, J., Agency for Science, Technology and Research
Liu, Z., Agency for Science, Technology and Research
Ang, E. L., Agency for Science, Technology and Research
Zhao, H., University of Illinois at Urbana-Champaign
DNA assembly forms the cornerstone of modern synthetic biology. Despite the numerous available methods, scarless multi-fragment assembly of large plasmids remains challenging. Furthermore, the upcoming wave in molecular biological automation demands a rethinking of how we perform DNA assembly. To streamline automation workflow and minimize operator intervention, a non-enzymatic assembly method is highly desirable. Here, we report the optimization and operationalization of a process called Twin-Primer Assembly (TPA), which is a method to assemble PCR-amplified fragments into a plasmid without the use of enzymes.

Each fragment used in TPA requires two PCR products, amplified separately using two sets of primers. One product has an overlap to the fragment that comes before and the other has an overlap to the fragment that comes after. The length of the overlap region is determined by the melting temperature (T_M) of the overlap region, and all overlaps used in an assembly are designed to have the same T_M. The two PCR products are mixed and fully denatured at high temperature before they are allowed to re-anneal slowly, generating intermediates that have overhangs on both sides. These intermediates are then allowed to hybridize at an elevated temperature to form a circular nicked plasmid. Upon *E. coli* transformation, the cells phosphorylate and ligate the nicks to create the final product plasmid.

The TPA non-enzymatic DNA assembly method is scarless and generally sequence independent. It is capable of assembling a 7 kb plasmid from 10 fragments at close to 80% fidelity, and a 31kb plasmid from 5 fragments at around 50% fidelity. Assembling more than 10 fragments is likely possible for constructing smaller plasmids, while assembling a plasmid larger than 31 kb is likely possible using fewer fragments. Even without the use of enzymes, the performance of TPA is on par with some of the best *in vitro* assembly methods currently available. TPA should be an invaluable addition to a synthetic biologistâ□™s toolbox.

High-Throughput Microfluidic Electroporation for Library Generation

- Conference: Synthetic Biology Engineering Evolution Design SEED
- Year: 2017
- Proceeding: 2017 Synthetic Biology: Engineering, Evolution & Design (SEED)
- Group: General Submissions
- Session:

  Session 10: Genetic Engineering Tools

- Time:

  Thursday, June 22, 2017 - 4:15pm-4:45pm

Authors:
Garcia, P. A., Massachusetts Institute of Technology
McClune, C., MIT
Laub, M., MIT
Buie, C. R., Massachusetts Institute of Technology
Voigt, C. A., Massachusetts Institute of Technology
Protein engineering has important applications that range from developing sensors in bacteria to producing antibodies for cancer therapeutics. Re-engineering protein function usually requires many amino acid changes, which results in libraries containing 10^{12}-10^{20} mutants. However, current genetic transformation tools limit bioengineers to making much smaller mutant libraries. This is relevant in applications where continuous transformation is desired such as in the creation of a library of mutants for drug discovery, metabolic engineering, and deep mutational scanning [1, 2]. In this presentation we will show recent efforts using a continuous flow microfluidic electroporation system to accelerate discovery in protein engineering.

Electroporation is a genetic engineering tool that uses pulsed electric fields for intracellular delivery of molecules such as drugs, proteins, DNA, or RNA and is traditionally performed in a single-batch cuvette system. A subset of the authors have developed a novel microfluidic electroporation platform for improving bacterial electrotransformation efficiency and throughput by up to 3 orders of magnitude [3]. The microfluidic system uses a non-uniform constriction to expose bacteria to a hydrodynamically controlled electric field that induces electroporation. The short residence time (~ms) within the constriction and fast flow rates employed (~mL/min) enables electrotransformation of sufficient bacteria to approach the required numbers needed for protein engineering.

In this study our efforts revolve around re-engineering bacterial two-component sensors, modulating the specificity of bacteriaâ□™s largest class of natural bio-sensors. We have designed a 7kb DNA construct that codes for Spectinomycin resistance and has a theoretical library size of $1x10^{16}$ mutants. *E. coli* cells in exponential growth phase (OD_{600}Â = 0.5) were harvested, washed 3x in 4Â°C 10% glycerol, and concentrated 200x prior to electrotransformation. The 1Â mL samples were transformed in a diverging microfluidic channels at 1.0Â mL/min, 1.5Â mL/min, and 2.0Â mL/min with processing times of 60Â s, 45Â s, or 30Â s, respectively. Preliminary quantification resulted in $5x10^8$ (1.0Â mL/min), $2.5x10^9$ (1.5Â mL/min), and $3.75x10^9$ (2.0Â mL/min) total transformants in comparison to the $2x10^7$ total transformants from a 50 Î¼L cuvette transformation. These results suggest that using a single microfluidic channel results in up to a 10x increase in transformation efficiency with an approximately 10-100x increase in throughput compared to cuvette based electroporation.

We envision this microfluidic platform as a tool to enable synthetic biologists to access the largest natural biosensors set in engineered systems. This study focuses on *E. coli* but could be easily applied to other microorganisms and adapted to parallel processing in order to get closer to the theoretical library size. Results of this study will broaden the complexity of engineered libraries that can be developed for protein engineering.

1. Smanski, M.J., et al., *Functional optimization of gene clusters by combinatorial design and assembly.* Nat Biotech, 2014. **32**(12): p. 1241-1249.

2. Fowler, D.M. and S. Fields, *Deep mutational scanning: a new style of protein science.* Nat Meth, 2014. **11**(8): p. 801-807.

3. Garcia, P.A., et al., *High efficiency hydrodynamic bacterial electrotransformation.* Lab Chip, 2017. **17**(3): p. 490-500.

Nanoparticles As Biomolecular Cargo Transporters in Plants and Plastids

- Conference: Synthetic Biology Engineering Evolution Design SEED
- Year: 2017
- Proceeding: 2017 Synthetic Biology: Engineering, Evolution & Design (SEED)
- Group: General Submissions
- Session:

  Session 10: Genetic Engineering Tools

- Time:

  Thursday, June 22, 2017 - 4:45pm-5:00pm

Authors:
Demirer, G. S., UC Berkeley
Landry, M., University of California Berkeley

The field of targeted plant genome editing has experienced exciting progress since the production of the first transgenic plants. Recent advances in targeted genome engineering development of zinc finger nucleases (ZFNs), TAL effector nucleases (TALENs), and CRISPR/Cas systemsâ“ allow modification of DNA sequences of interest, once inside living cells.[1] To introduce genome engineering tools into plant cells and plastids, several methods have been established, such as Agrobacterium or polyethylene glycol (PEG) mediated gene delivery, electroporation, biolistic particle gun, and microinjection. However, current methods are limited by host range limitations, low transfection efficiency, toxicity, and in most cases, transfection limited to protoplast cultures where the cell wall is removed, and not to the mature whole plant.[2] We explore the use of high-aspect ratio nanomaterials for efficient through-cell-wall gene transfer into all desired genotypes of any plant species.

Nanomaterials, with distinct and tunable physical properties (size, shape, and surface charge), effectively deliver biological cargoes to traditionally hard-to-access intracellular locations, such as walled plant cells. We have previously shown that certain nanomaterial formulations can penetrate the mature plant cell, and also the chloroplast.[3,4] Here, we present recent results on the passive delivery of nanoparticles, loaded with plasmid DNA, into walled plant cells and protoplasts. In particular, we examine the delivery mediated by single-walled carbon nanotubes (SWCNT) functionalized with GFP encoding plasmids, and the resulting GFP expression in the transfected cells of mature arugula plants. Our work provides a promising tool for generic, targeted, and passive transport of biomolecules (genes, proteins, etc.) into mature plant cells for targeted genome editing and plastid engineering.

References

1. Li, J.-F. et al. Multiplex and homologous recombination-mediated genome editing in Arabidopsis and Nicotiana benthamiana using guide RNA and Cas9. Nature biotechnology 31, 688-691 (2013).
2. Potrykus, I. Gene Transfer to Plants: Assessment of Published Approaches and Results. Annual Review of Plant Physiology and Plant Molecular Biology 42, 205-225, doi:10.1146/annurev.pp.42.060191.001225 (1991).
3. Giraldo, J. P. et al. Plant nanobionics approach to augment photosynthesis and biochemical sensing. Nat Mater 13, 400-408, doi:10.1038/nmat3890
4. Wong, M. H. et al. Lipid Exchange Envelope Penetration (LEEP) of Nanoparticles for Plant Engineering: A Universal Localization Mechanism. Nano Letters 16, 1161-1172, doi:10.1021/acs.nanolett.5b04467 (2016).

Adding to the Genome: Accuate Targeting and Effecient Insertion of Long ssDNA

- Conference: Synthetic Biology Engineering Evolution Design SEED
- Year: 2017
- Proceeding: 2017 Synthetic Biology: Engineering, Evolution & Design (SEED)
- Group: General Submissions
- Session:

  Session 10: Genetic Engineering Tools

- Time:

  Thursday, June 22, 2017 - 5:00pm-5:15pm

Authors:
Clore, A., Integrated DNA Technologies
The development of CRISPR-mediated genome targeting is one of the most disruptive technologies of this decade, however the usefulness of this system alone is limited to the inactivation of genetic information. Homology mediated repair allows the creation and repair of genetic information at the CRISPR break, and is most efficient when single stranded template is used.

Using novel technologies we demonstrate the creation of ssDNA kilobases in size, and the insertion of such DNA into targeted break sites with a variety of mammalian cell lines and animal models, often with as little as 50 bases of homology. The efficiency of this method is augmented by the use of a chemically modified crRNA/tracrRNA/Cas9. Data summarizing the methods of design, transfection, and screening using this system will be presented.

Use of Synthetic Sgrna to Improve CRISPR Editing Efficiency

- Conference: Synthetic Biology Engineering Evolution Design SEED
- Year: 2017
- Proceeding: 2017 Synthetic Biology: Engineering, Evolution & Design (SEED)
- Group: General Submissions
- Session:

  Session 10: Genetic Engineering Tools

- Time:

  Thursday, June 22, 2017 - 5:15pm-5:30pm

Authors:
Holden, K.

CRISPR has made genome editing accessible for a wide range of cell-types. However, obtaining consistent editing efficiencies remains a challenge. Synthego has developed novel RNA synthesis technology to produce 100-mer sgRNA for CRISPR at a practical scale and price. We demonstrate, through a collaborative effort with key researchers utilizing CRISPR, that Synthego sgRNA produces consistent and superior genome editing in a variety of cell types including adherent mammalian cell lines, primary T-cells, iPSCs and model organism embryos. This technology will help to enable succesful and consistent genome editing for both basic research models and for sensitive primary human cells to be used for therapeutic applications.

Synthetic Biology Platforms for Natural Product Biosynthesis and Discovery

- Conference: Synthetic Biology Engineering Evolution Design SEED
- Year: 2017
- Proceeding: 2017 Synthetic Biology: Engineering, Evolution & Design (SEED)
- Group: General Submissions
- Session:

  Keynote 4

- Time:

  Friday, June 23, 2017 - 8:30am-9:30am

Authors:
Smolke, C., Stanford University
Plants are a rich source of unique scaffolds, including 25% of natural-product-derived drugs. However, the discovery, synthesis, and overall material supply chains for sourcing plant natural products and their derivatives remain ad hoc, biased, and tedious. While microbial biosynthesis presents compelling alternatives to traditional approaches based on extraction from natural plant hosts, many challenges exist in the reconstruction of plant specialized metabolic pathways in microbial hosts. My laboratory has developed approaches to address the challenges that arise in the reconstruction of complex biosynthesis schemes, including spatial engineering strategies to direct the activities and specificities of pathway enzymes and recoding strategies to address folding, processing, and stability issues that may arise with the expression of plant enzymes in heterologous microbial hosts. We have utilized these strategies to develop yeast-based production platforms for an important class of plant alkaloids, the benzylisoquinoline alkaloids, including the medicinal opioids. These synthetic biology platforms will lead to transformative advances in natural product discovery, drug development, and production.

Computational Design of Novel Enzymes and Metabolic Pathways for Industrial Biotechnology

- Conference: Synthetic Biology Engineering Evolution Design SEED
- Year: 2017
- Proceeding: 2017 Synthetic Biology: Engineering, Evolution & Design (SEED)
- Group: General Submissions
- Session:

  Session 11: Mathematical Modeling and Design

- Time:

  Friday, June 23, 2017 - 10:00am-10:30am

Authors:
Zanghellini, A., Arzeda
Ban, Y. E. A., Arzeda
Galdzicki, M., Arzeda Corp
Toofanny, R., Arzeda
Medley, K.
Palunas, K.
Grabs, D., Arzeda
Althoff, E., Arzeda Corp
Sauro, H. M., University of Washington
Our ability to design cell factories to produce valuable chemicals requires the â□œrecombinationâ□□ not only of existing but also designer enzymes into novel metabolic pathways to achieve entirely new metabolic function. This poses two distinct challenges that need to be solved in an integrated way if we ambition to fully deliver on the promise of synthetic biology. The first challenge deals with the design, at scale, of novel enzymes with high level of activities for reactions not known to be catalyzed in nature. To this end, Arzeda has developed and scaled high-throughput computational methodologies to exploit natural enzyme latent catalytic promiscuity and rapidly design new biocatalysts. A successful industrial application of this technology will be presented.

The second challenge is that of finding optimal ways to arrange natural and designed enzymes to biosynthetize a small-molecule of interest. To illustrate our progress towards solving this challenge, we will discuss a novel computational tool that Arzeda has developed for the automated design of novel biosynthetic pathways. Inspired by retrosynthetic methods in organic chemistry, our software draws on databases of known natural enzymatic reactions as well as reactions that can be catalyzed by designed enzymes to exhaustively enumerate biosynthetic routes from a set of desired metabolites down to any small molecule of interest. Pathways are ranked based on thermodynamic feasibility and other metrics. We will discuss applications of industrial relevance in the field of fine and bulk chemicals.

A Tug-of-War Mechanism for Pattern Formation in a Genetic Network

- Conference: Synthetic Biology Engineering Evolution Design SEED
- Year: 2017
- Proceeding: 2017 Synthetic Biology: Engineering, Evolution & Design (SEED)
- Group: General Submissions
- Session:

  Session 11: Mathematical Modeling and Design

- Time:

  Friday, June 23, 2017 - 10:30am-11:00am

Authors:
Gomez, M. M., UC Berkeley
Arcak, M., University of California, Berkeley

Synthesizing spatial patterns with genetic networks is an ongoing challenge in synthetic biology. A successful demonstration of pattern formation would imply a better understanding of systems in the natural world and advance applications in synthetic biology. In developmental systems, transient patterning may suffice in order to imprint instructions for long-term development. In this paper we show that transient but persistent patterns can emerge from a realizable synthetic gene network based on a toggle switch. We show that a bistable system incorporating diffusible molecules can generate patterns that resemble Turing patterns but are distinctly different in the underlying mechanism: diffusion of mutually inhibiting molecules creates a prolonged â□œtug-of-warâ□□

between patches of cells at opposing bistable states. The patterns are observed to be transient but longer wavelength patterns persist for extended periods of time. Analysis of a representative small scale model implies the eigenvalues of the persistent modes are just above the threshold of stability. The results are verified through simulation of biologically relevant models.

A Synthetic Biology Workflow Using Sbol

- Conference: Synthetic Biology Engineering Evolution Design SEED
- Year: 2017
- Proceeding: 2017 Synthetic Biology: Engineering, Evolution & Design (SEED)
- Group: General Submissions
- Session:

  Session 11: Mathematical Modeling and Design

- Time:

  Friday, June 23, 2017 - 11:30am-12:00pm

Authors:
Myers, C. J., University of Utah
McLaughlin, J., Newcastle University
Misirli, G., Newcastle University
Nguyen, T., University of Utah
Samineni, M., University of Utah
Zhang, M., University of Utah
Zundel, Z., University of Utah
Wipat, A., Newcastle University

A synthetic biology workflow requires genetic design repositories to organize collections of parts, sequence-level design and visualization tools to compose genetic parts into complete constructs, and genetic circuit design tools to construct, model, and simulate complete systems enabling the evaluation of alternative design choices. Data standards, such as the Synthetic Biology Open Language (SBOL), are critical to enabling the exchange of information within such a workflow, allowing a diversity of repositories and software tools to be seamlessly connected. This talk describes one such workflow that leverages the growing ecosystem of software tools that support SBOL to describe genetic designs, and the mature ecosystem of tools that support the Systems Biology Markup Language (SBML) to model these designs. In particular, this presentation will demonstrate a workflow using tools including SynBioHub, SBOLDesigner, and iBioSim. SynBioHub (http://synbiohub.org) is a database designed for storing synthetic biology designs captured using the SBOL data model, and it provides both a RESTful API for computational access and a user-friendly Web-based frontend. SBOLDesigner is a sequence editor that allows the designer to fetch parts from a SynBioHub repository and compose them to construct larger designs that can be submitted back into the repository. Finally, iBioSim is a genetic modeling, analysis, and design tool that provides a means to construct complete circuit designs in SBOL and pair them with SBML models for these designs that can be simulated and analyzed using a variety of techniques. Both SBOLDesigner and iBioSim also support uploading these larger system designs back to the SynBioHub repository. Finally, this talk will demonstrate how this workflow can be utilized to produce SBOL for the ACS Synthetic Biology journal, which has recently recommended the use of SBOL in their publications.

A Comprehensive Characterization of the Nitroreductase Superfamily

- Conference: Synthetic Biology Engineering Evolution Design SEED
- Year: 2017
- Proceeding: 2017 Synthetic Biology: Engineering, Evolution & Design (SEED)
- Group: Poster Session
- Session:

  Confirmed Posters

Authors:
Morales, D., Michael Smith Laboratories (UBC)
Copp, J., University of British Columbia
Akiva, E., University of California San Francisco
Babbitt, P., University of California San Francisco
Tokuriki, N., University of British Columbia

A genetic toggle switch is a circuit that possesses two stable opposite states of gene expression. Many synthetic circuits that are applied to biotechnology are based on toggle switches, such as biosensors. However, extrinsic and intrinsic noise can cause random switching and generate a variety of gene expression states. To make circuits more robust, we propose the creation of a consensus toggle switch that will have the ability to amplify induction signals among neighboring cells and achieve a consensus in the population. We will create the consensus by employing quorum sensing in our circuits. Quorum sensing (QS) is the mechanism through which bacteria communicate with each other, by producing signaling molecules that can affect gene expression in a density-dependent way. We hypothesize that synthetic toggles that coordinate their behavior with QS can better perform tasks that are difficult for uncoordinated systems. Here, we constructed 4 versions of consensus toggles to understand how the presence/absence of positive feedback loops in quorum-sensing genes and direct/indirect repression affect toggle behavior. In our designs, each possible state of the toggle is reported by one color: yellow (YFP) or blue (CFP). We find that the dynamics of consensus toggle switches depend on their regulatory topologies. Topologies that contain indirect repression show a peak of YFP expression followed by a shutdown, while topologies with direct repression do not shut down. Furthermore, topologies with indirect repression do not show distinguishable CFP expression between induced and uninduced samples, while topologies with direct repression do. Topologies that lack extra feedback loops show less variability, while topologies that contain these extra feedback loops are more variable. These results enable the selection of a topology that works better for the creation of a consensus toggle. In the microscope, the toggle with direct repression and without extra feedback loops showed faster switch compared to a conventional toggle. These results are encouraging, because a consensus toggle has the potential of generating more sensitive and robust detection of signals than regular toggles. And to further apply this circuit, such as for the detection of diseases/infections in the mammalian gastrointestinal tract or for the surveillance and bioremediation of environments, it is desired that the circuit shows sensitivity and robustness.

A CRISPR/Cas9 Genome Engineering Toolbox for Systems Metabolic Engineering of Yeast

- Conference: Synthetic Biology Engineering Evolution Design SEED
- Year: 2017
- Proceeding: 2017 Synthetic Biology: Engineering, Evolution & Design (SEED)
- Group: Poster Session
- Session:

  Confirmed Posters

Authors:
Verwaal, R., DSM Biotechnology Center
Buiting-Wiessenhaan, N., DSM Biotechnology Center
Vonk, B., DSM Biotechnology Center
Roubos, J. A., DSM Biotechnology Center
Plants are a rich source of unique scaffolds, including 25% of natural-product-derived drugs. However, the discovery, synthesis, and overall material supply chains for sourcing plant natural products and their derivatives remain ad hoc, biased, and tedious. While microbial biosynthesis presents compelling alternatives to traditional approaches based on extraction from natural plant hosts, many challenges exist in the reconstruction of plant specialized metabolic pathways in microbial hosts. My laboratory has developed approaches to address the challenges that arise in the reconstruction of complex biosynthesis schemes, including spatial engineering strategies to direct the activities and specificities of pathway enzymes and recoding strategies to address folding, processing, and stability issues that may arise with the expression of plant enzymes in heterologous microbial hosts. We have utilized these strategies to develop yeast-based production platforms for an important class of plant alkaloids, the benzylisoquinoline alkaloids, including the medicinal opioids. These synthetic biology platforms will lead to transformative advances in natural product discovery, drug development, and production.

A Genetic Switch for Stable, Long-Term Fermentative Production of Anabolic Products in Yeast

- Conference: Synthetic Biology Engineering Evolution Design SEED
- Year: 2017
- Proceeding: 2017 Synthetic Biology: Engineering, Evolution & Design (SEED)
- Group: Poster Session
- Session:

  Confirmed Posters

Authors:
Chua, P., Amyris, Inc.
Meadows, A., Amyris, Inc.
Jiang, H., Amyris, Inc.
Plants are a rich source of unique scaffolds, including 25% of natural-product-derived drugs. However, the discovery, synthesis, and overall material supply chains for sourcing plant natural products and their derivatives remain ad hoc, biased, and tedious. While microbial biosynthesis presents compelling alternatives to traditional approaches based on extraction from natural plant hosts, many challenges exist in the reconstruction of plant specialized metabolic pathways in microbial hosts. My laboratory has developed approaches to address the challenges that arise in the reconstruction of complex biosynthesis schemes, including spatial engineering strategies to direct the activities and specificities of pathway enzymes and recoding strategies to address folding, processing, and stability issues that may arise with the expression of plant enzymes in heterologous microbial hosts. We have utilized these strategies to develop yeast-based production platforms for an important class of plant alkaloids, the benzylisoquinoline alkaloids, including the medicinal opioids. These synthetic biology platforms will lead to transformative advances in natural product discovery, drug development, and production.

A Massively Parallel Assay for Studying 5' UTR Dependent Translation Regulation in Mammalian Cells

- Conference: Synthetic Biology Engineering Evolution Design SEED
- Year: 2017
- Proceeding: 2017 Synthetic Biology: Engineering, Evolution & Design (SEED)
- Group: Poster Session
- Session:

  Confirmed Posters

Authors:
Wang, B.
Seelig, G., University of Washington
Sample, P., University of Washington
Recent studies into the interplay between the microbiome and host have extensively implicated short-chain fatty acid (SCFA) fermentation products as contributors to both intra and extra intestinal health. In the context of the gut, butyrate (a four carbon SCFA) serves as the primary energy source for epithelial colonocytes, drives regulatory T cell differentiation and is therefore essential for the maintenance of colonic mucosal health. On a systemic level, butyrate may influence insulin sensitivity, cholesterol synthesis and autoimmunity. Typically, butyrate is produced by the action of slow-growing Gram positive organisms â□“ often Clostridial in origin â□“ during the fermentation of cellulose and complex carbohydrates. To counteract the depletion of butyrate in the gut, we iteratively engineered *Escherichia coli* Nissle 1917 to produce butyrate. The new butyrate-producing Nissle was capable of millimolar level production of butyrate comparable with the levels produced by enteric Clostridial species. The circuitry was placed under anaerobic control to support induction upon entry into the gastrointestinal track. Finally, proof-of-concept *in vivo* experimentation showed production of butyrate in the mouse gut. This engineered strain may be useful for the treatment of a variety of gastrointestinal and systemic conditions associated with SCFA deficiency.

A Model-Driven Approach to Improving Cooperativity and Reducing Toxicity of dCas9 Logic Circuits

Conference: Synthetic Biology Engineering Evolution Design SEED

Year: 2017

Proceeding: 2017 Synthetic Biology: Engineering, Evolution & Design (SEED)

Group: Poster Session

Session: Confirmed Posters

Authors:

Anderson, D. A., Massachusetts Institute of Technology
Zhang, S., MIT
Voigt, C. A., Massachusetts Institute of Technology

Small molecule-inducible transcription factors, or sensors, enable temporal control of toxic genes, provide the inputs forÂ in vivoÂ computation, and allow a host to sense and respond to its environment. In this ongoing work we endeavor to improve the characteristics of several sensors using a mixture of rational engineering and directed evolution. We have improved the dynamic range, sensitivity, and/or specificity of 18 sensors. As an example of the utility of these sensors, we have used five optimized sensors to independently control the expression of five components of a metabolic pathway. By varying the amount of each inducer, we can easily and rapidly optimize the expression profile of the pathway without the need to construct large libraries. To facilitate the adoption of multiplex, sensor-based genetic control, we have created a stable cell-line capable of responding to 11 inputs, thus enabling the modular control of multiple genetic outputs with minimal optimization.

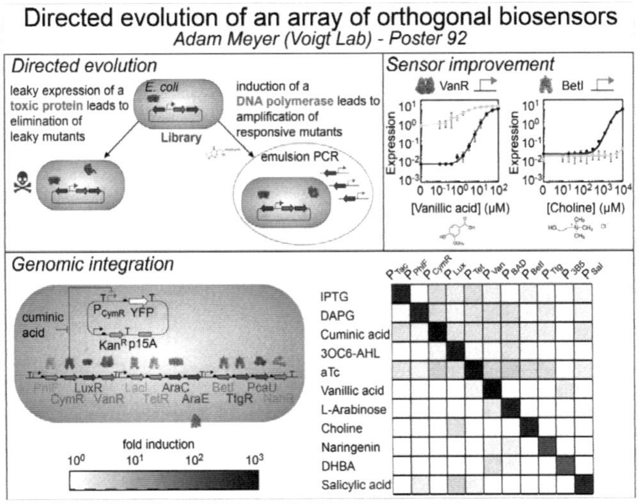

A Next-Generation Multi-Omics Approach to Dissect Cellular and Molecular Reprogramming

- Conference: Synthetic Biology Engineering Evolution Design SEED
- Year: 2017
- Proceeding: 2017 Synthetic Biology: Engineering, Evolution & Design (SEED)
- Group: Poster Session
- Session:

  Confirmed Posters

Authors:
Mandal, K., National Institute of Immunology
Sen Sharma, S., National Institute of Immunology
Yadav, G., National Institute of Plant Genome Research
Majumdar, S. S., National Institute of Immunology

Protein glycosylation, the post-translational attachment of complex oligosaccharides (glycans), is the most abundant polypeptide modification in eukaryotes, and is integrally involved in human biology and disease. Changes in glycosylation have been shown to modulate the pharmacokinetics and potency of protein therapeutics and vaccines, making them critical quality attributes for biotechnology applications. However, the study of glycans in native systems (glycobiology) and the intentional manipulation of protein glycosylation patterns (glycoengineering) remain limited by a shortage of methods to characterize glycosylation enzyme activities and a reliance on endogenous glycosylation systems which can produce highly variable glycosylation patterns and constrain choices of production hosts, protein trafficking, and possible glycoforms. The structural complexity of enzymes involved in initial modification of polypeptides with glycans and the vast sequence spaces which must be surveyed to determine their specificities have impeded efforts to implement synthetic glycosylation systems in heterologous production hosts, such as *Escherichia coli*. New tools are required to rapidly express polypeptide modifying glycosyltransferases and determine design rules for engineering protein glycosylation sites which can be efficiently and specifically modified with a glycan of interest.

Here we describe an *in vitro* platform for high-throughput expression and characterization of glycosylation enzymes using *E. coli* Cell-Free Protein Synthesis (CFPS) and Self-Assembled Monolayers for Desorption Ionization Mass Spectrometry (SAMDI-MS). This workflow allowed us to produce >800 μg/mL of a cytoplasmic N-linked glycosyltransferase (NGT) *in vitro* and determine its peptide acceptor and sugar donor specificities at unprecedented depth and throughput with ~10,000 unique reactions conditions and ~3,000 unique peptide substrates. We used this dataset to develop a small, robust acceptor sequence motif (GlycTag) to direct the efficient installation of N-linked glycans onto the internal loops of heterologous protein substrates *in vitro* and in the cytoplasm of living *E. coli*. We found that SAMDI-MS observations at the peptide level accurately predicted modification trends in the context of whole proteins. We placed GlycTags in three structurally diverse target proteins, including the Fc region of a human antibody, and found that these sequences were modified at 3-5 fold greater efficiency compared to wild-type glycosylation sequence motifs. Notably, this N-linked glycosylation system does not require protein transport across cellular membranes or the use of membrane bound components, making it particularly attractive for the development of synthetic glycosylation systems.

This work provides a disruptive technique for glycosyltransferase characterization and provides a new set of tools towards emerging bacterial and *in vitro* glycosylation platforms. Ultimately, these glycoengineering efforts will enable a deeper understanding of glycan structure and function and help to bring about a new generation of rationally designed glycoprotein therapeutics, vaccines, and diagnostics.

A Quorum Sensing-Based Lateral Inhibition System to Generate Contrasting Patterns

- Conference: Synthetic Biology Engineering Evolution Design SEED
- Year: 2017
- Proceeding: 2017 Synthetic Biology: Engineering, Evolution & Design (SEED)
- Group: Poster Session
- Session:

  Confirmed Posters

Authors:
Tei, M., UC Berkeley
Perkins, M. L., University of California, Berkeley
Hsia, J., University of Washington, Seattle
Arkin, A., University of California, Berkeley
Arcak, M., University of California, Berkeley

Inhibitory interactions play an important role in the design and function of both natural and engineered biological systems. Lateral inhibition, which occurs when spatially adjacent units inhibit each other, facilitates pattern formation in multicellular organisms during development; for example, somitogenesis via the Notch/Delta pathway in metazoans. Here, we combine quorum sensing and compartmentalization to implement lateral inhibition between neighboring bacterial colonies connected by a channel. Unlike the Notch/Delta pathway, our system does not employ contact-mediated inhibition, as there are no known contact-dependent signaling pathways in bacteria that do not involve growth inhibition. Instead, we harness intercellular communication via diffusible molecules to drive divergence in reporter gene expression between neighboring colonies. Contrasting patterns emerge when, at steady state, colonies producing high levels of reporter are immediately adjacent to colonies producing low levels of reporter. We develop a mathematical model for a series of interconnected, mutually inhibitory compartments to predict quantitatively the parameters and spatial configurations that allow contrasting patterns to emerge. We then engineer two bacterial strains that repress each otherâ□™s transcription of a reporter gene and demonstrate that contrasting patterns do indeed arise. Establishing patterning in a simple bacterial system will permit us to explore the mathematical principles underlying biological development in a controlled manner.

A Quorum Sensing-Based Lateral Inhibition System Generates Contrasting Patterns (Poster #144)

Melinda L. Perkins,[1*†] Mika Tei,[2†] Justin Hsia,[1,3] Adam Arkin,[2] Murat Arcak[1]

[1] Department of Electrical Engineering, University of California, Berkeley, CA, USA, [2] Department of Bioengineering, University of California, Berkeley, CA, USA, [3] Department of Computer Science and Engineering, University of Washington, Seattle, WA, USA, *Presenting author (mindylp@berkeley.edu), †Equal contributors

lateral inhibition between bacterial colonies

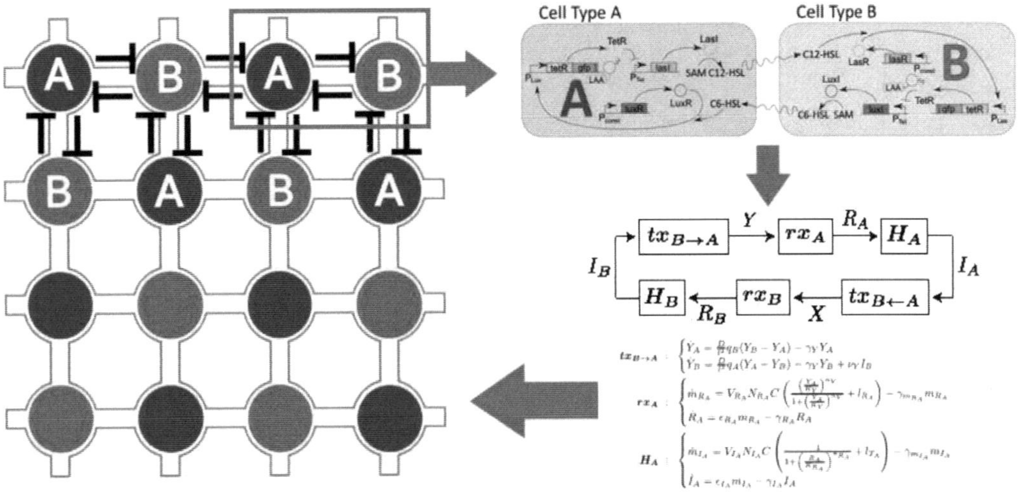

bistability produces contrasting patterns

A Quorum-Busting Approach to Microbiome Engineering

- Conference: Synthetic Biology Engineering Evolution Design SEED
- Year: 2017
- Proceeding: 2017 Synthetic Biology: Engineering, Evolution & Design (SEED)
- Group: Poster Session
- Session:

  Confirmed Posters

Authors:
Treviño González, M. T., Universidad Autónoma de Nuevo León
Morones, J. R., Universidad Autónoma de Nuevo León
Sepúlveda Guzmán, S., Universidad Autónoma de Nuevo León

The availability of cheap and scalable gene synthesis techniques will open up the exploration of sequence-function relationships at unprecedented scales. Attempts to reduce the costs of gene synthesis have focused on how to assemble microarray derived oligo (OLS) pools into longer length genes. Here we present a new approach capable of assembling hundreds of genes simultaneously from a complex oligo pool by localizing only the fragments required for each individual gene assembly to barcoded beads before isolating the assembly reaction inside an emulsion droplet. Multiplexed reactions of 384 genes, each assembled from five oligos, have costs approaching $1 per 500 bp gene. We apply this to assemble a thousand phylogenetically diverse orthologs of phosphopantetheine adenylyltransferase (PPAT) representing over 500 kbp of coding sequence and use a sequencing based barcode readout to carry out a complementation assay in E. coli, highlighting how this technique can be applied to determine functional relationships of long sequences in high-throughput at low-cost.

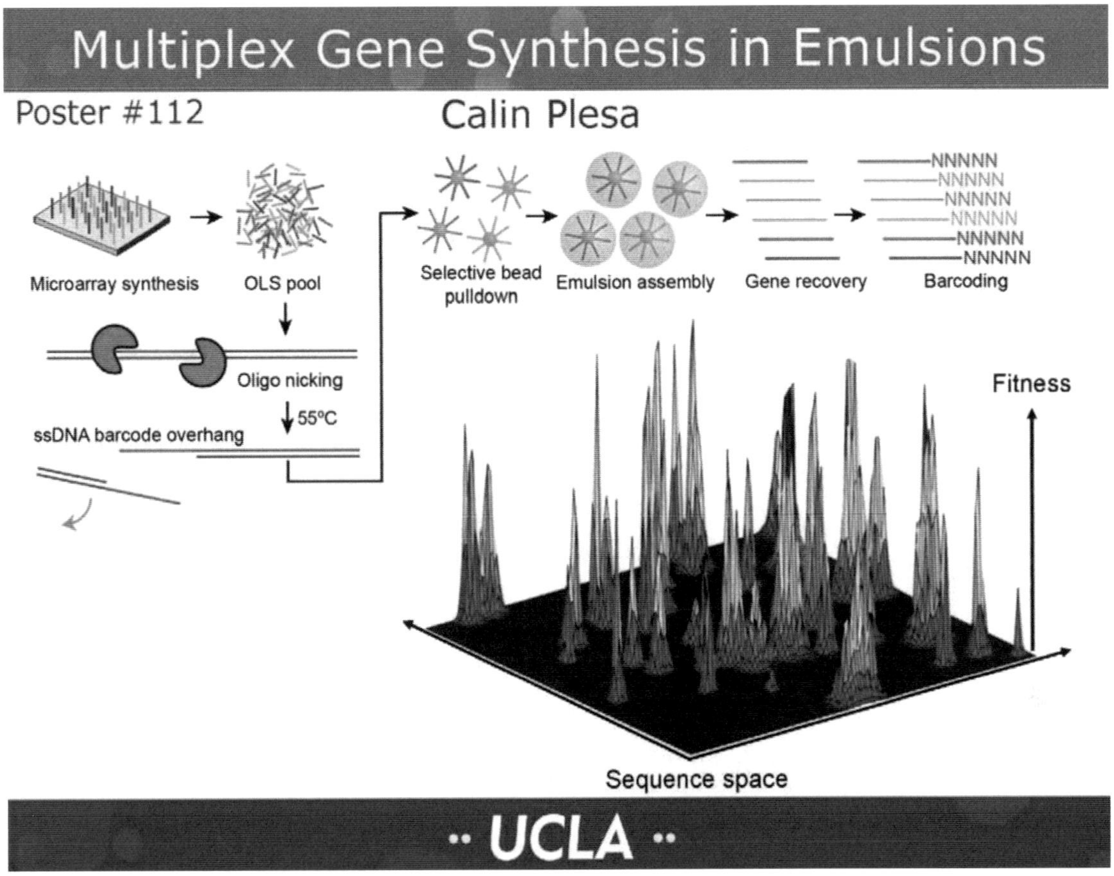
Multiplex Gene Synthesis in Emulsions
Poster #112
Calin Plesa
Microarray synthesis
OLS pool
Oligo nicking
55°C
ssDNA barcode overhang
Selective bead pulldown
Emulsion assembly
Gene recovery
Barcoding
NNNNN
Fitness
Sequence space
UCLA

A rational approach for bio-production of recombinant proteins in mammalian cell systems

- Conference: Synthetic Biology Engineering Evolution Design SEED
- Year: 2017
- Proceeding: 2017 Synthetic Biology: Engineering, Evolution & Design (SEED)
- Group: Poster Session
- Session:

  Confirmed Posters

Authors:
Altamura, R.
Doshi, J.
Benenson, Y., Swiss Federal Institute of Technology (ETH Zürich)

Bacteria lack compartmentalization and therefore rely on subcellular organization to temporally and spatially localize major cellular processes. As one of the major structures in the cell, the nucleoid possesses positional cues that help direct important functions such as DNA replication and cell division. While some of the factors that contribute to maintaining subcellular organization are characterized, many of the underlying determinants are still unknown. In *Bacillus subtilis*, there remains a class of putative DNA-binding proteins currently uncharacterized, and I hypothesize that some of these proteins use DNA to localize functions important in maintaining subcellular organization. Since many uncharacterized genes lack easily discernable knockout phenotypes, I utilized a gene misexpression approach to identify DNA-binding proteins capable of perturbing some aspect of cellular organization. The misexpression library employed comprises over 800 strains, each containing an uncharacterized gene under the control of an inducible promoter on the *B. subtilis* chromosome. Genes for 98 putative DNA-binding proteins were misexpressed and analyzed for their ability to grow in the presence of inducer. Of these strains, nine exhibited reduced growth and 17 strains did not grow. When analyzed by fluorescence microscopy, 19 of these strains could be associated with a distinct phenotype, including defects in the structure and placement of the chromosome, abnormal cell division, and loss of cell shape. To further characterize strains with the most severe phenotypes we performed RNA-seq experiments to determine if gene misexpression alters the transcription profile. We identified four genes that do not affect transcription, suggesting that the misexpression phenotype is not due to transcriptional regulation. We hypothesize that this class of DNA-binding proteins act as positional cues that help maintain subcellular organization.

•

A Serine Integrase-Based 'Landing Pad' System for Chromosomal Integrations of Heterologous Genes in E. coli

- Conference: Synthetic Biology Engineering Evolution Design SEED
- Year: 2017
- Proceeding: 2017 Synthetic Biology: Engineering, Evolution & Design (SEED)
- Group: Poster Session
- Session:

  Confirmed Posters

Authors:
Clubbe, R.

Recent studies into the interplay between the microbiome and host have extensively implicated short-chain fatty acid (SCFA) fermentation products as contributors to both intra and extra intestinal health. In the context of the gut, butyrate (a four carbon SCFA) serves as the primary energy source for epithelial colonocytes, drives regulatory T cell differentiation and is therefore essential for the maintenance of colonic mucosal health. On a systemic level, butyrate may influence insulin sensitivity, cholesterol synthesis and autoimmunity. Typically, butyrate is produced by the action of slow-growing Gram positive organisms â□" often Clostridial in origin â□" during the fermentation of cellulose and complex carbohydrates. To counteract the depletion of butyrate in the gut, we iteratively engineered *Escherichia coli* Nissle 1917 to produce butyrate. The new butyrate-producing Nissle was capable of millimolar level production of butyrate comparable with the levels produced by enteric Clostridial species. The circuitry was placed under anaerobic control to support induction upon entry into the gastrointestinal track. Finally, proof-of-concept *in vivo* experimentation showed production of butyrate in the mouse gut. This engineered strain may be useful for the treatment of a variety of gastrointestinal and systemic conditions associated with SCFA deficiency.

A Toolkit of Inducible Degron-Tagged dCas9 Effectors for Multi-Directional Drug-Tunable Control of Synthetic Biology Applications

- Conference: Synthetic Biology Engineering Evolution Design SEED
- Year: 2017
- Proceeding: 2017 Synthetic Biology: Engineering, Evolution & Design (SEED)
- Group: Poster Session
- Session:

  Confirmed Posters

Authors:
Kleinjan, D. J., University of Edinburgh
Rosser, S. J., University of Edinburgh
Wardrope, C., University of Edinburgh
Sou, S. N., University of Edinburgh
A wide variety of amino acyl tRNA synthetases (aaRSs) have had their substrate specificities altered through protein engineering to create mutant aaRSs that can charge tRNA with non-canonical amino acids in cells. While the activities of these mutant aaRSs can be dynamically controlled in cells using conditional promoters, aaRS can also be synthesized using non-natural amino acids as they accumulate in cells, which in some cases could negatively impact their ability to metabolically label other proteins. In addition, our ability to quickly switch aaRS-mediated metabolic labeling on and off remains limited because we have not yet discovered ways to directly control aaRS activity post-translationally in response to environmental changes. To develop prototype aaRSs with activities that are dependent upon direct binding to a chemical, we have engineered a first generation of ligand-responsive aaRSs. For these efforts, we targeted a mutant *Escherichia coli* methionyl tRNA synthetase (MetRS), which was previously developed to charge tRNA with azidonorleucine (Anl) in *Escherichia coli*. We first used a combinatorial approach to discover split MetRS whose fragments cooperatively function when each fragment is fused to a pair of proteins that interact. We then examined whether the activity of split MetRS can be regulated by either fusing the fragments to a pair of proteins whose interaction is stabilized by ligand binding or by fusing the fragments to the termini of a single protein domain that exhibits a ligand-dependent conformational change. We have found that both approaches can be used to regulate Anl labeling of newly synthesized proteins in a ligand-dependent manner. When MetRS fragments were fused to FKBP12 and the rapamycin binding domain of mTOR, metabolic labeling was significantly enhanced in growth medium containing rapamycin,

which stabilizes the FKBP12-mTOR complex. In addition, fusion of MetRS fragments to the termini of the ligand-binding domain of the estrogen receptor yielded a protein whose metabolic labeling was significantly enhanced when 4-hydroxytamoxifen was added to the growth medium. Ongoing efforts are focused on assessing whether the ligand-dependent metabolic labeling mediated by our engineered proteins can be switched on faster than labeling achieved using transcriptional regulation, improving the metabolic labeling efficiencies that can be achieved with our prototype aaRS switches, and investigating the diversity of ligands that can be used to directly regulate MetRS activity. The protein switches arising from these studies are expected to be useful for extending our control over protein metabolic labeling with Anl by allowing for fast regulation of this labeling using post-translational reactions. Furthermore, this approach can be applied to metabolic labeling with additional non-natural amino acids by extending our designs to structurally-related aaRS with distinct substrate specificities. Ligand-responsive aaRSs are expected to have applications in the production of protein biomaterials, proteomic studies, and protein pharmaceutical biosynthesis.

A Versatile Toolbox for the Control of Protein Levels Using N?-Acetyl-Lysine Dependent Amber Suppression

- Conference: Synthetic Biology Engineering Evolution Design SEED
- Year: 2017
- Proceeding: 2017 Synthetic Biology: Engineering, Evolution & Design (SEED)
- Group: Poster Session
- Session:

  Confirmed Posters

Authors:
Volkwein, W., Ludwig-Maximilians Universität München
Maier, C., Ludwig-Maximilians Universität München
Krafczyk, R., Ludwig-Maximilians Universität München
Jung, K., Ludwig-Maximilians Universität München
Lassak, J., Ludwig-Maximilians Universität München
Mosquito-borne diseases such as Zika, dengue fever and Chikungunya are major and increasing international public health concerns. The primary vector of these viral diseases is *Aedes aegypti* and current control measures against this mosquito are proving inadequate or problematic. An innovative and effective control technique for *Aedes aegypti* involves the release of self-limiting mosquitoes. These mosquitoes are genetically engineered to be homozygous for a self-limiting gene under control of an on/off tetracycline switch. When released into the wild, any offspring of the transgenic males will inherit one copy of the transgene. In the absence of tetracycline antidote, the transgene will prevent the mosquito from developing to adulthood, thus suppressing the population with repeated releases. Oxitecâ□™s self-limiting transgenic insect technology has also been applied successfully to develop population control measures for major crop pests such as the Mediterranean fruit fly (Medfly) and the diamondback moth. In a variation on the self-limiting genetics used in *Aedes* mosquitoes, genetic sexing strains that express the self-limiting gene in a sex-specific manner are used to produce male-only cohorts of these crop pests for release in control programmes. These and other developments in Oxitecâ□™s genetic engineering programme to enable biological control of major insect pests will be discussed. The Oxitec transgenic mosquito method has dramatically reduced *Aedes aegypti* populations in trials around the world and is now being used under operational conditions in the Cayman Islands and Brazil to control *Aedes aegypti*. Results of these programmes will be presented.

•

Acoustic Reporter Genes for Non-Invasive Deep Tissue Imaging of Engineered Cells

- Conference: Synthetic Biology Engineering Evolution Design SEED
- Year: 2017
- Proceeding: 2017 Synthetic Biology: Engineering, Evolution & Design (SEED)
- Group: Poster Session
- Session:

  Confirmed Posters

Authors:
Farhadi, A., California Institute of Technology
Bourdeau, R. W., California Institute of Technology
Ho, G., Caltech
Lee-Gosselin, A., California Institute of Technology
Lakshmanan, A., California Institute of Technology
Kumar, S., California Institute of Technology
Nety, S. P., California Institute of Technology
Shapiro, M. G., California Institute of Technology
Identifying, understanding, and ultimately engineering the complex set of interacting genes responsible for any cellular phenotype is extremely difficult and traditionally requires years of study. For example, optimizing metabolic flux to a native or non-native small molecule or increasing the diversity of carbon sources that a microbe can use are well-studied topics, and yet very few novel compounds have made it through the high risk path to commercialization. To address this problem, Zymergen has developed a powerful platform for automated high-throughput microbial strain development that is focused on improving the economics of large-scale fermentation processes and accelerating the industrialization of novel bio-products. The platform has been designed to be flexible, highly reliable, and host-agnostic. With its high throughput capability Zymergen tests thousands of design ideas across many library types to identify those that improve performance. Automation makes Zymergenâ□™s strain improvement process both precise and repeatable. Using this platform, Zymergen takes a data-driven approach to biology, capturing as much data as possible and using heuristically-guided search processes that embrace the complexity of biology.

Active Degradation Allows for Increased Information Transmission Rate in Genetic Control

- Conference: Synthetic Biology Engineering Evolution Design SEED
- Year: 2017
- Proceeding: 2017 Synthetic Biology: Engineering, Evolution & Design (SEED)
- Group: Poster Session
- Session:

  Confirmed Posters

Authors:
Rossi, N. A., Boston Univeristy
Dunlop, M. J., N/A

In genetic regulation, time scale matters. Many organisms exist in changing, uncertain environments in which they are required to respond quickly to stress. They must do this without compromising the fidelity of the response. In this research, we use the multiple antibiotic resistance activator MarA from *Escherichia coli* as a case study in how active degradation of a master regulator balances tradeoffs in information transmission clarity and response rates. We use a combination of experimental, analytical and computation methods to consider how well a signal from MarA propagates to downstream genes and how effectively MarA coordinates the response between these downstream genes. If time scale is not important, a slowly changing regulator maximizes the information transmission and downstream coordination. However, over shorter time-scales, a quickly changing actively degraded regulator is ideal.Â This research demonstrates the utility of active degradation in genetic regulation and how it is essential to consider the time-scales of the stress to which the regulatory elements are adapted.

Poster 121

Active Degradation Allows for Increased Information Transmission Rate in Genetic Control

Nicholas Rossi[1], Aleksandra Walczak[2], Thierry Mora[2], Mary Dunlop[3]
(1)Molecular Biology, Cell Biology and Biochemistry Program, Boston University (2)Laboratoire de Physique Theorique, Ecole Normale Supeneure,(3)Department of Biomedical Engineering, Boston University

Active degradation of a master regulator

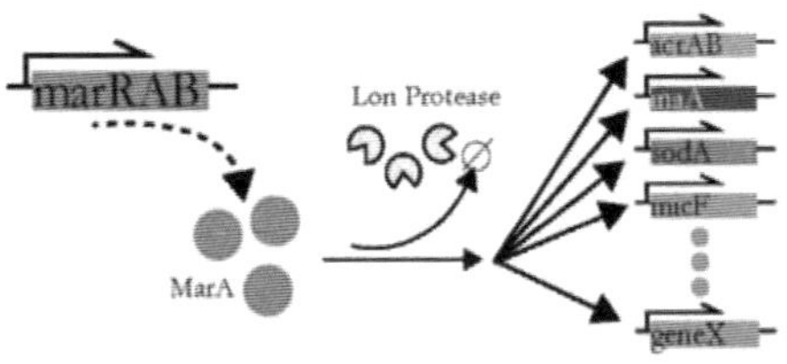

Accelerated coordination of multiple downstream gene targets

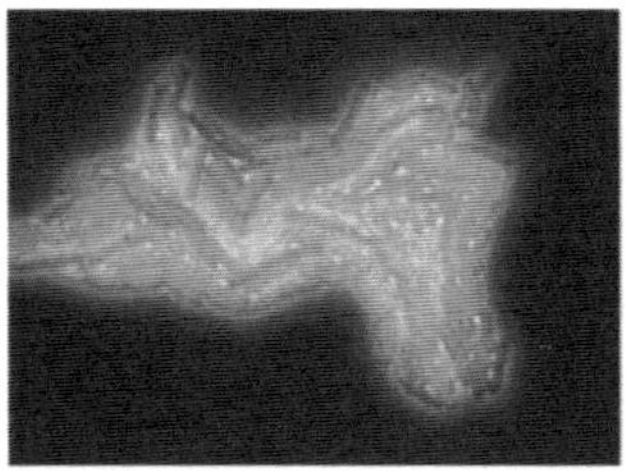

Rapid accumulation of diversity within isogenic bacterial populations

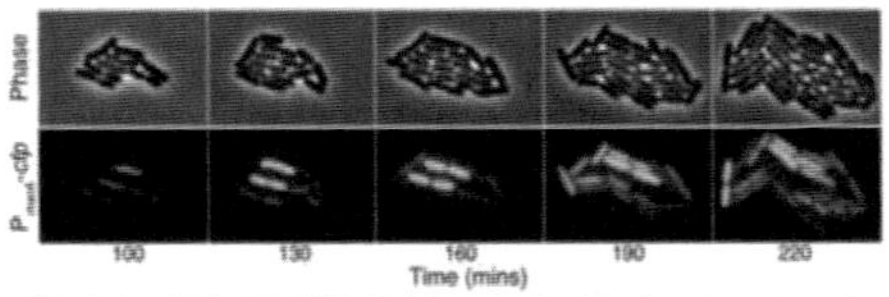

El Meouche, Imane, Yik Siu, and Mary J. Dunlop. "Stochastic expression of a multiple antibiotic resistance activator confers transient resistance in single cells." Scientific reports 6 (2016).

Tools from synthetic biology can help us understand this process

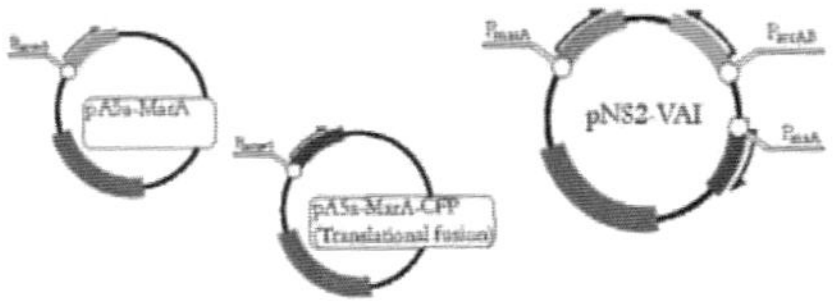

Digital version of poster with videos, animations and interactives: www.rossidata.com/SEED2017

•

Adaptive Landscapes and Molecular Basis for the Functional Innovation of an Organophosphate Hydrolysing Enzyme

- Conference: Synthetic Biology Engineering Evolution Design SEED
- Year: 2017
- Proceeding: 2017 Synthetic Biology: Engineering, Evolution & Design (SEED)
- Group: Poster Session
- Session:

  Confirmed Posters

Authors:
Yang, G., University of British Columbia
Tokuriki, N., University of British Columbia
Jackson, C. J., Australian National University
Baier, F.
Bornberg-Bauer, E.
Dohmen, E.
Anderson, D.

Synthetic Biology aims to implement new cellular features through the design of original genetic parts and systems. A wide variety of circuit architectures have been built successfully, proving the ability of synthetic genetic parts to predictably interact with each other within a single bacterial cell [4]. The next stage of synthetic biology will involve the *de novo* engineering of self-organizing multicellular systems. One of the main methods found in nature for creating complex spatially distributed systems is cellular differentiation through asymmetric cell division (ACD), whereby stem cells spawn off irreversibly differentiated daughters. The presence of different cell lineages within an eukaryotic tissue allows for the partitioning of metabolic and structural tasks. Furthermore, if one or both of the daughter cells becomes incapable of dividing, controlling asymmetric division probability can lead to tissue size and shape control [5]. In complex eukaryotic organisms ACD is typically brought about by the differential expression of key genes within each cell type. Here, we engineered a novel method for generating ACD in *Escherichia coli* and regulating its frequency within the population. This work provides a platform for the construction of complex synthetic tissues and organisms from bacterial cells that differentiate during development of the population.

Current methods for creating differentiated cell types in synthetic bacteria utilize gene regulatory networks that allow for bistable gene expression [2, 1]. While such systems generate bimodal populations, they are very sensitive to both intrinsic and extrinsic noise sources that can randomly force transitions between states [3]. We use a novel method for creating terminally differentiated synthetic bacterial cells. Specifically, we refactored DNA shuttling mechanisms from *Caulobacter crescentus*

into *E. coli* to target plasmid DNA into one and only one daughter cell upon division. Thus, the two daughter cells will not just have different gene expression states, but will be irreversibly genetically distinct. Our findings will lead to synthetic "tissues" that can be used in complex settings.

References

[1] R. G. Egbert and E. Klavins. Fine-tuning gene networks using simple sequence repeats.*Proceedings of the National Academy of Sciences*, 109(42):16817â□“16822, 2012.
[2] T. S. Gardner, C. R. Cantor, and J. J. Collins. Construction of a genetic toggle switch in *Escherichia coli*. *Nature*, 403(6767):339â□“342, 2000.
[3] C. Gupta, J. M. LÃ³pez, W. Ott, K. JosiÄ‡, and M. R. Bennett. Transcriptional delay stabilizes bistable gene networks. *Physical review letters*, 111(5):058104, 2013.
[4] A. S. Khalil and J. J. Collins. Synthetic biology: applications come of age. *Nature Reviews*

Genetics, 11(5):367â□“379, 2010.
[5] A. D. Lander, K. K. Gokoffski, F. Y. Wan, Q. Nie, and A. L. Calof. Cell lineages and the logic of proliferative control. *PLoS Biol*, 7(1):e1000015, 2009.

Amplifying Riboswitch Signal Output Using Cellular Wiring

- Conference: Synthetic Biology Engineering Evolution Design SEED
- Year: 2017
- Proceeding: 2017 Synthetic Biology: Engineering, Evolution & Design (SEED)
- Group: Poster Session
- Session:

  Confirmed Posters

Authors:
Bennett, A., Air Force Research Laboratory
Goodson, M., Air Force Research Laboratory
Kelley-Loughnane, N., Air Force Research Laboratory
Harbaugh, S., Air Force Research Laboratory

Biocatalyst discovery is integral to bioecomony development, enabling design of robust and scalable bioprocesses that are able to compete with the resource intensive petrochemical industry. Uncultivated microbial communities within natural and engineered ecosystems provide a near-infinite reservoir of genomic diversity and metabolic potential that can be harnessed for biocatalyst discovery[1]. To bridge the cultivation gap, functional metagenomic screens have been developed to recover active genes directly from environmental samples[2]. Here, we describe a pipeline for recovery of biomass-deconstructing biocatalysts sourced from pulp and paper mill sludge metagenomes to develop consolidated bioprocesses that convert waste streams into value-added products. High molecular weight DNA was extracted from sludge and used to construct a fosmid library containing 15,000 clones using the copy control system in EPI300â„¢-T1 RÂ *E. coliÂ* based on established robotic methods[3]. Extracted DNA was also used in whole genome shotgun sequencing to compare the metabolic potential of the sludge community with fosmid screening outcomes using MetaPathways[4]Â with specific emphasis on carbohydrate-active enzymes (CAZymes). In total, 32,232 ORFâ□™s hits in CAZy database were obtained including glycoside hydrolases, glycosyl transferases, and carbohydrate binding module families. The fosmid library was screened for glycosidase hydrolase activities using a pool of sensitive fluorogenic glycosides of 6-chloro-4-methylumbelliferoneÂ enabling rapid recovery of Î²-cellobiosidase, mannosidase and xylosidase activitiesÂ 5. A total of 744 clones capable of converting pooled substrates were recovered indicating an extremely high hit rate (1 hit per 43 clones). Downstream characterization including complete fosmid sequencing and subcloning results to identify specific active genes or gene cassettes is currently in progress. These results reinforce the notion that screening libraries sourced from environments naturally enriched for lignocellulosic biomass deconstruction increases screening efficieny and chances of recovering biocatalysts optimized to perform under specific environmental conditions. The combination of robotic library production and pooled screening using enriched DNA sources for lignocellulosic biomass deconstruction provides an example of high throughput experimental design extensible to a wide range of screening targets.

1. Mewis, K., Armstrong, Z., Song, Y. C., Baldwin, S. A., Withers, S. G., & Hallam, S. J. (2013). Biomining active cellulases from a mining bioremediation system. *Journal of biotechnology*, *167*(4), 462-471.
2. Nyyssönen, M., Tran, H. M., Karaoz, U., Weihe, C., Hadi, M. Z., Martiny, J. B., ... & Brodie, E. L. (2013). Coupled high-throughput functional screening and next generation sequencing for identification of plant polymer decomposing enzymes in metagenomic libraries. *Frontiers in microbiology*, *4*, 282.
3. Mewis, K., Taupp, M., & Hallam, S. J. (2011). A high throughput screen for biomining cellulase activity from metagenomic libraries. *JoVE (Journal of Visualized Experiments)*, (48), e2461-e2461.
4. Hanson, N. W., Konwar, K. M., Wu, S. J., & Hallam, S. J. (2014, May). MetaPathways v2. 0: A master-worker model for environmental Pathway/Genome Database construction on grids and clouds. In *Computational Intelligence in Bioinformatics and Computational Biology, 2014 IEEE Conference on*(pp. 1-7). IEEE.
5. Chen, Hong-Ming, et al. "Synthesis and evaluation of a series of 6-chloro-4-methylumbelliferyl glycosides as fluorogenic reagents for screening metagenomic libraries for glycosidase activity." Carbohydrate research 421 (2016): 33-39.

•

An Accelerated Evolution Approach to Produce Next Generation Yeast Strains for Protein Discovery and Production

- Conference: Synthetic Biology Engineering Evolution Design SEED
- Year: 2017
- Proceeding: 2017 Synthetic Biology: Engineering, Evolution & Design (SEED)
- Group: Poster Session
- Session:

  Confirmed Posters

Authors:
Comyn, S.
Mathieu, Y.
Celedon, J.
Brumer, H.
Bohlmann, J.
Tokuriki, N.
Mayor, T.
Plants are a rich source of unique scaffolds, including 25% of natural-product-derived drugs. However, the discovery, synthesis, and overall material supply chains for sourcing plant natural products and their derivatives remain ad hoc, biased, and tedious. While microbial biosynthesis presents compelling alternatives to traditional approaches based on extraction from natural plant hosts, many challenges exist in the reconstruction of plant specialized metabolic pathways in microbial hosts. My laboratory has developed approaches to address the challenges that arise in the reconstruction of complex biosynthesis schemes, including spatial engineering strategies to direct the activities and specificities of pathway enzymes and recoding strategies to address folding, processing, and stability issues that may arise with the expression of plant enzymes in heterologous microbial hosts. We have utilized these strategies to develop yeast-based production platforms for an important class of plant alkaloids, the benzylisoquinoline alkaloids, including the medicinal opioids. These synthetic biology platforms will lead to transformative advances in natural product discovery, drug development, and production.

An Automated Flow-Based/Digital Microfluidic Platform with on-Chip Optical Detection System for High-Throughput Gene Editing Processes

- Conference: Synthetic Biology Engineering Evolution Design SEED
- Year: 2017
- Proceeding: 2017 Synthetic Biology: Engineering, Evolution & Design (SEED)
- Group: Poster Session
- Session:

  Confirmed Posters

Authors:
Iwai, K., Joint BioEnergy Institute
Gach, P., Joint BioEnergy Institute
Kim, P. W., Sandia National Laboratories
Raje, M., Joint BioEnergy Institute
Heinemann, J., Joint BioEnergy Institute
A. Duncombe, T., UCSF/UC Berkeley Joint Graduate Group in Bioengineering, University of California
Deng, K., Joint BioEnergy Institute
Northen, T., Joint BioEnergy Institute
Hillson, N. J., Joint BioEnergy Institute
Adams, P. D., Joint BioEnergy Institute
Singh, A. K., Joint Bioenergy Institute

The development of CRISPR-mediated genome targeting is one of the most disruptive technologies of this decade, however the usefulness of this system alone is limited to the inactivation of genetic information. Homology mediated repair allows the creation and repair of genetic information at the CRISPR break, and is most efficient when single stranded template is used.

Using novel technologies we demonstrate the creation of ssDNA kilobases in size, and the insertion of such DNA into targeted break sites with a variety of mammalian cell lines and animal models, often with as little as 50 bases of homology. The efficiency of this method is augmented by the use of a chemically modified crRNA/tracrRNA/Cas9. Data summarizing the methods of design, transfection, and screening using this system will be presented.

An Engineered Pyrrolysyl tRNA Synthetase for In Vivo Measurement of Sense Codon Reassignment in E. coli

- Conference: Synthetic Biology Engineering Evolution Design SEED
- Year: 2017
- Proceeding: 2017 Synthetic Biology: Engineering, Evolution & Design (SEED)
- Group: Poster Session
- Session:

  Confirmed Posters

Authors:
Schwark, D., Colorado State University
Schmitt, M. A., Colorado State University
Fisk, J. D., Colorado State University

Plants are a rich source of unique scaffolds, including 25% of natural-product-derived drugs. However, the discovery, synthesis, and overall material supply chains for sourcing plant natural products and their derivatives remain ad hoc, biased, and tedious. While microbial biosynthesis presents compelling alternatives to traditional approaches based on extraction from natural plant hosts, many challenges exist in the reconstruction of plant specialized metabolic pathways in microbial hosts. My laboratory has developed approaches to address the challenges that arise in the reconstruction of complex biosynthesis schemes, including spatial engineering strategies to direct the activities and specificities of pathway enzymes and recoding strategies to address folding, processing, and stability issues that may arise with the expression of plant enzymes in heterologous microbial hosts. We have utilized these strategies to develop yeast-based production platforms for an important class of plant alkaloids, the benzylisoquinoline alkaloids, including the medicinal opioids. These synthetic biology platforms will lead to transformative advances in natural product discovery, drug development, and production.

Ancestral Reconstruction of LacI/GalR Transcriptional Regulators

- Conference: Synthetic Biology Engineering Evolution Design SEED
- Year: 2017
- Proceeding: 2017 Synthetic Biology: Engineering, Evolution & Design (SEED)
- Group: Poster Session
- Session:

  Confirmed Posters

Authors:
Spence, M., Australian National University
Kaczmarski, J., Australian National University
Jackson, C. J., Australian National University
Recent studies into the interplay between the microbiome and host have extensively implicated short-chain fatty acid (SCFA) fermentation products as contributors to both intra and extra intestinal health. In the context of the gut, butyrate (a four carbon SCFA) serves as the primary energy source for epithelial colonocytes, drives regulatory T cell differentiation and is therefore essential for the maintenance of colonic mucosal health. On a systemic level, butyrate may influence insulin sensitivity, cholesterol synthesis and autoimmunity. Typically, butyrate is produced by the action of slow-growing Gram positive organisms â€“ often Clostridial in origin â€“ during the fermentation of cellulose and complex carbohydrates. To counteract the depletion of butyrate in the gut, we iteratively engineered *Escherichia coli* Nissle 1917 to produce butyrate. The new butyrate-producing Nissle was capable of millimolar level production of butyrate comparable with the levels produced by enteric Clostridial species. The circuitry was placed under anaerobic control to support induction upon entry into the gastrointestinal track. Finally, proof-of-concept *in vivo* experimentation showed production of butyrate in the mouse gut. This engineered strain may be useful for the treatment of a variety of gastrointestinal and systemic conditions associated with SCFA deficiency.

An E coli whole Genome Mutant Library Enables Rapid Selection for Improved Fitness Under Diverse Stresses

- Conference: Synthetic Biology Engineering Evolution Design SEED
- Year: 2017
- Proceeding: 2017 Synthetic Biology: Engineering, Evolution & Design (SEED)
- Group: Poster Session
- Session:

  Confirmed Posters

Authors:
Choudhury, A., University of Colorado, Boulder
Wang, Z., Tianjin University
Gill, D. R. T., University of Colorado
Kaar, J. L., University of Colorado Boulder

Synthetic Biology aims to implement new cellular features through the design of original genetic parts and systems. A wide variety of circuit architectures have been built successfully, proving the ability of synthetic genetic parts to predictably interact with each other within a single bacterial cell [4]. The next stage of synthetic biology will involve the *de novo* engineering of self-organizing multicellular systems. One of the main methods found in nature for creating complex spatially distributed systems is cellular differentiation through asymmetric cell division (ACD), whereby stem cells spawn off irreversibly differentiated daughters. The presence of different cell lineages within an eukaryotic tissue allows for the partitioning of metabolic and structural tasks. Furthermore, if one or both of the daughter cells becomes incapable of dividing, controlling asymmetric division probability can lead to tissue size and shape control [5]. In complex eukaryotic organisms ACD is typically brought about by the differential expression of key genes within each cell type. Here, we engineered a novel method for generating ACD in *Escherichia coli* and regulating its frequency within the population. This work provides a platform for the construction of complex synthetic tissues and organisms from bacterial cells that differentiate during development of the population.

Current methods for creating differentiated cell types in synthetic bacteria utilize gene regulatory networks that allow for bistable gene expression [2, 1]. While such systems generate bimodal populations, they are very sensitive to both intrinsic and extrinsic noise sources that can randomly force transitions between states [3]. We use a novel method for creating terminally differentiated synthetic bacterial cells. Specifically, we refactored DNA shuttling mechanisms from *Caulobacter crescentus*

into *E. coli* to target plasmid DNA into one and only one daughter cell upon division. Thus, the two daughter cells will not just have different gene expression states, but will be irreversibly genetically distinct. Our findings will lead to synthetic "tissues" that can be used in complex settings.

References

[1] R. G. Egbert and E. Klavins. Fine-tuning gene networks using simple sequence repeats.*Proceedings of the National Academy of Sciences*, 109(42):16817â“16822, 2012.
[2] T. S. Gardner, C. R. Cantor, and J. J. Collins. Construction of a genetic toggle switch in *Escherichia coli*. *Nature*, 403(6767):339â“342, 2000.
[3] C. Gupta, J. M. LÃ³pez, W. Ott, K. JosiÄ‡, and M. R. Bennett. Transcriptional delay stabilizes bistable gene networks. *Physical review letters*, 111(5):058104, 2013.
[4] A. S. Khalil and J. J. Collins. Synthetic biology: applications come of age. *Nature Reviews*

Genetics, 11(5):367â“379, 2010.
[5] A. D. Lander, K. K. Gokoffski, F. Y. Wan, Q. Nie, and A. L. Calof. Cell lineages and the logic of proliferative control. *PLoS Biol*, 7(1):e1000015, 2009.

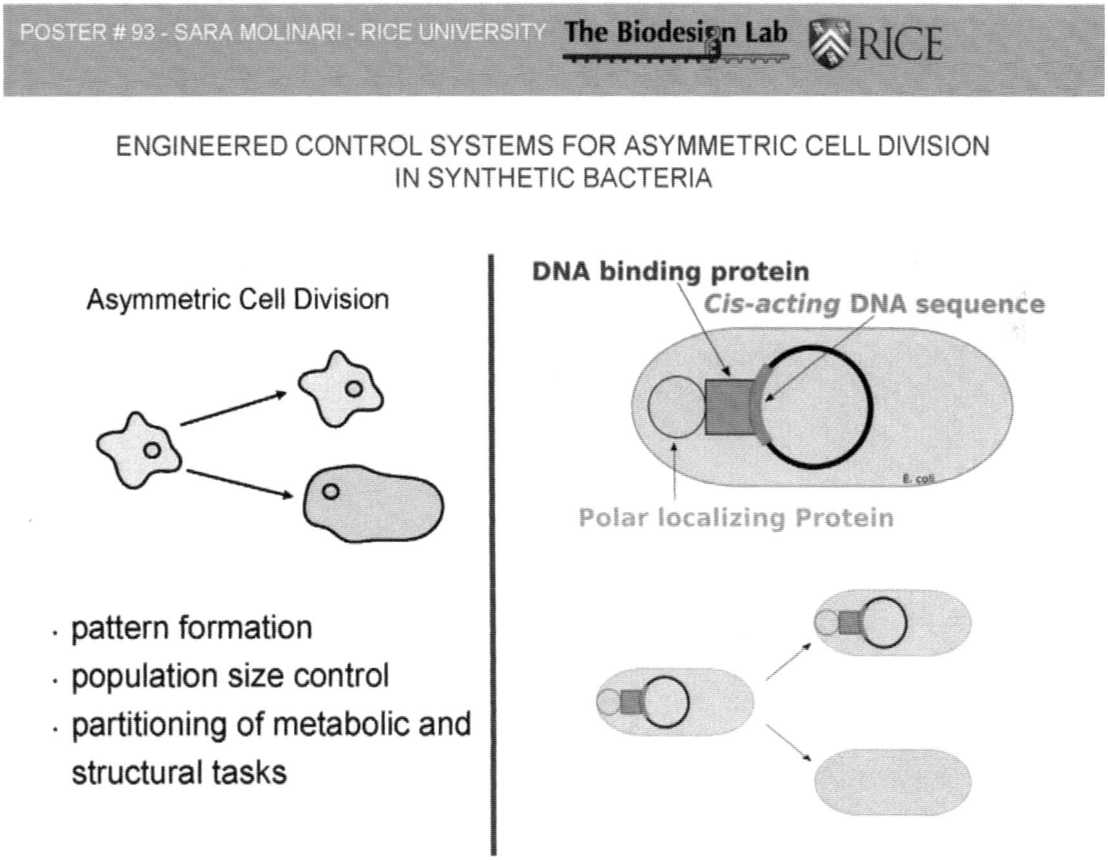

Applying Synthetic Biology to Synthesize Functional Materials

- Conference: Synthetic Biology Engineering Evolution Design SEED
- Year: 2017
- Proceeding: 2017 Synthetic Biology: Engineering, Evolution & Design (SEED)
- Group: Poster Session
- Session:

  Confirmed Posters

Authors:
Wallace, A., Massachusetts Institute of Technology
Voigt, C. A., Massachusetts Institute of Technology

Development of advanced functional materials with highly organized nano- and microstructures are of increasing demand across the medical, optical, energy, and mechanical fields. The functionality of materials is dependent on the relationship between structure and composition, demonstrating the importance of controlling molecular assembly throughout a material. In this work, we are applying several of natureâ□™s biological mechanisms for building structurally complex materials to engineer new biomaterials that display a wide range of functional properties. First, we are exploring whether post-translational modifications to the silica mineralizing R5 peptide from diatoms can be used to control silica morphology at the submicron scales. Second, we are growing melanin on R5 to create R5-melanin embedded silica materials. Finally, we are combining R5 with proteins from magnetotactic bacteria to synthesize silica-coated magnetic nanoparticles. By integrating these approaches for controlling material structure and composition, we aim to demonstrate the feasibility of using proteins to synthesize multifunctional materials.

Applying Synthetic Biology to Synthesize Functional Materials

Andrea Wallace
Poster #154

Project Goals

1. Develop a modular, protein-based materials synthesis platform.
2. Introduce post-translational modifications to peptides from diatoms to synthesize silica-based materials with functional properties and complex structures.

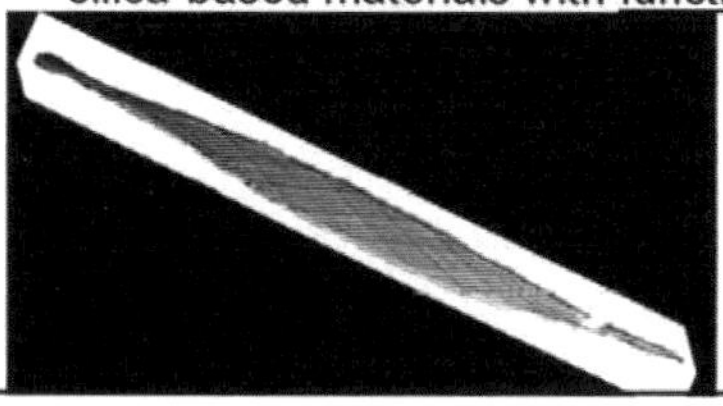

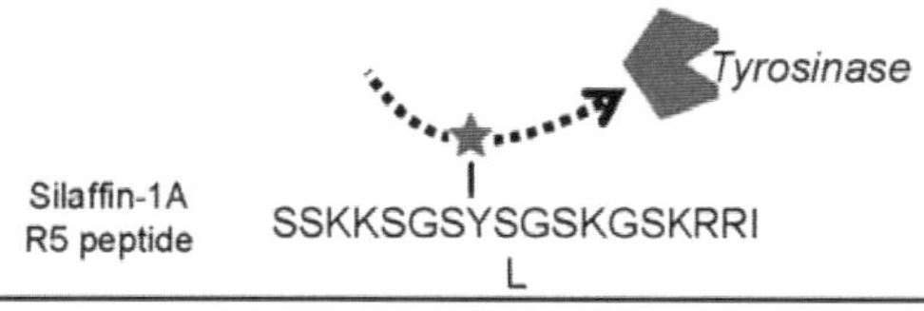

Results

Silica precipitation and characterization

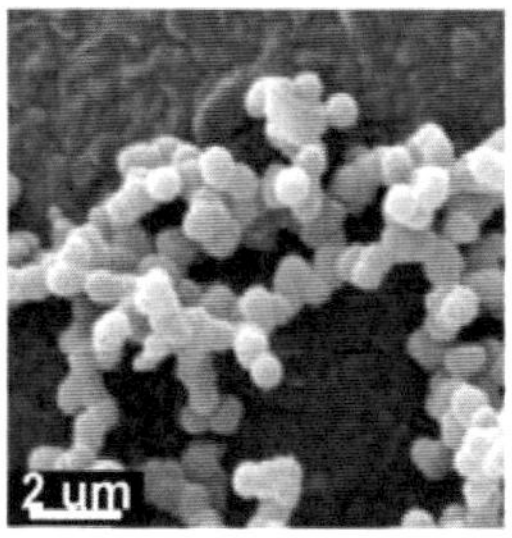

Melanin-like polymers embedded in silica

Silica-coated magnetosomes for microwave absorption

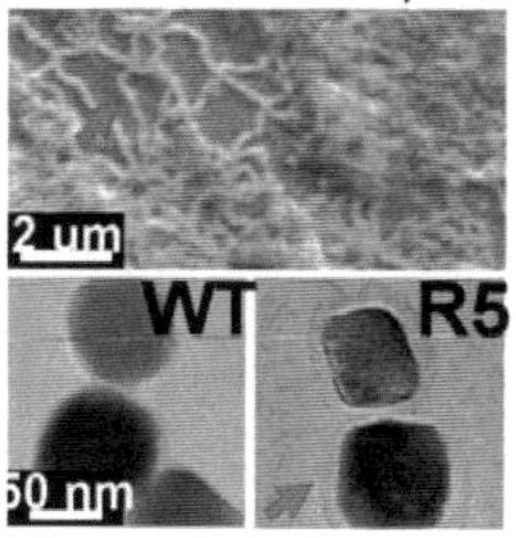

Assemblx: A User-Friendly Toolkit for Rapid and Reliable Multi-Gene Assemblies

- Conference: Synthetic Biology Engineering Evolution Design SEED
- Year: 2017
- Proceeding: 2017 Synthetic Biology: Engineering, Evolution & Design (SEED)
- Group: Poster Session
- Session:

  Confirmed Posters

Authors:
Machens, F., University of Potsdam
Hochrein, L., University of Potsdam
Messerschmidt, K., University of Potsdam
Mueller-Roeber, B., University of Potsdam
Schulz, K., University of Potsdam
Gremmels, J.

Identifying, understanding, and ultimately engineering the complex set of interacting genes responsible for any cellular phenotype is extremely difficult and traditionally requires years of study. For example, optimizing metabolic flux to a native or non-native small molecule or increasing the diversity of carbon sources that a microbe can use are well-studied topics, and yet very few novel compounds have made it through the high risk path to commercialization. To address this problem, Zymergen has developed a powerful platform for automated high-throughput microbial strain development that is focused on improving the economics of large-scale fermentation processes and accelerating the industrialization of novel bio-products. The platform has been designed to be flexible, highly reliable, and host-agnostic. With its high throughput capability Zymergen tests thousands of design ideas across many library types to identify those that improve performance. Automation makes Zymergenâ□™s strain improvement process both precise and repeatable. Using this platform, Zymergen takes a data-driven approach to biology, capturing as much data as possible and using heuristically-guided search processes that embrace the complexity of biology.

AT-Rich Gene Synthesis

- Conference: Synthetic Biology Engineering Evolution Design SEED
- Year: 2017
- Proceeding: 2017 Synthetic Biology: Engineering, Evolution & Design (SEED)
- Group: Poster Session
- Session:

  Confirmed Posters

Authors:
Walsh, M., MIT Lincoln Laboratory
Niles, J., Massachusetts Institute of Technology
Carr, P. A., MIT Lincoln Laboratory
Guido, N., Gen9
Plants are a rich source of unique scaffolds, including 25% of natural-product-derived drugs. However, the discovery, synthesis, and overall material supply chains for sourcing plant natural products and their derivatives remain ad hoc, biased, and tedious. While microbial biosynthesis presents compelling alternatives to traditional approaches based on extraction from natural plant hosts, many challenges exist in the reconstruction of plant specialized metabolic pathways in microbial hosts. My laboratory has developed approaches to address the challenges that arise in the reconstruction of complex biosynthesis schemes, including spatial engineering strategies to direct the activities and specificities of pathway enzymes and recoding strategies to address folding, processing, and stability issues that may arise with the expression of plant enzymes in heterologous microbial hosts. We have utilized these strategies to develop yeast-based production platforms for an important class of plant alkaloids, the benzylisoquinoline alkaloids, including the medicinal opioids. These synthetic biology platforms will lead to transformative advances in natural product discovery, drug development, and production.

Attempts to Develop an Enzyme Converting R-2,3-Dihydroxyisovalerate to 2-Ketoisovalerate with the Aid of Computational Protein Design

- Conference: Synthetic Biology Engineering Evolution Design SEED
- Year: 2017
- Proceeding: 2017 Synthetic Biology: Engineering, Evolution & Design (SEED)
- Group: Poster Session
- Session:

  Confirmed Posters

Authors:
Oki, K.
Mayo, S. L., California Institute of Technology
Lee, F. S.

Plants are a rich source of unique scaffolds, including 25% of natural-product-derived drugs. However, the discovery, synthesis, and overall material supply chains for sourcing plant natural products and their derivatives remain ad hoc, biased, and tedious. While microbial biosynthesis presents compelling alternatives to traditional approaches based on extraction from natural plant hosts, many challenges exist in the reconstruction of plant specialized metabolic pathways in microbial hosts. My laboratory has developed approaches to address the challenges that arise in the reconstruction of complex biosynthesis schemes, including spatial engineering strategies to direct the activities and specificities of pathway enzymes and recoding strategies to address folding, processing, and stability issues that may arise with the expression of plant enzymes in heterologous microbial hosts. We have utilized these strategies to develop yeast-based production platforms for an important class of plant alkaloids, the benzylisoquinoline alkaloids, including the medicinal opioids. These synthetic biology platforms will lead to transformative advances in natural product discovery, drug development, and production.

Automated Technologies for Synthetic Biology Applications

- Conference: Synthetic Biology Engineering Evolution Design SEED
- Year: 2017
- Proceeding: 2017 Synthetic Biology: Engineering, Evolution & Design (SEED)
- Group: Poster Session
- Session:

  Confirmed Posters

Authors:
Shih, S. C. C., Center For Applied Synthetic Biology
Sinha, H., Concordia University
Perry, J., Concordia University
Moazami, E., Concordia University

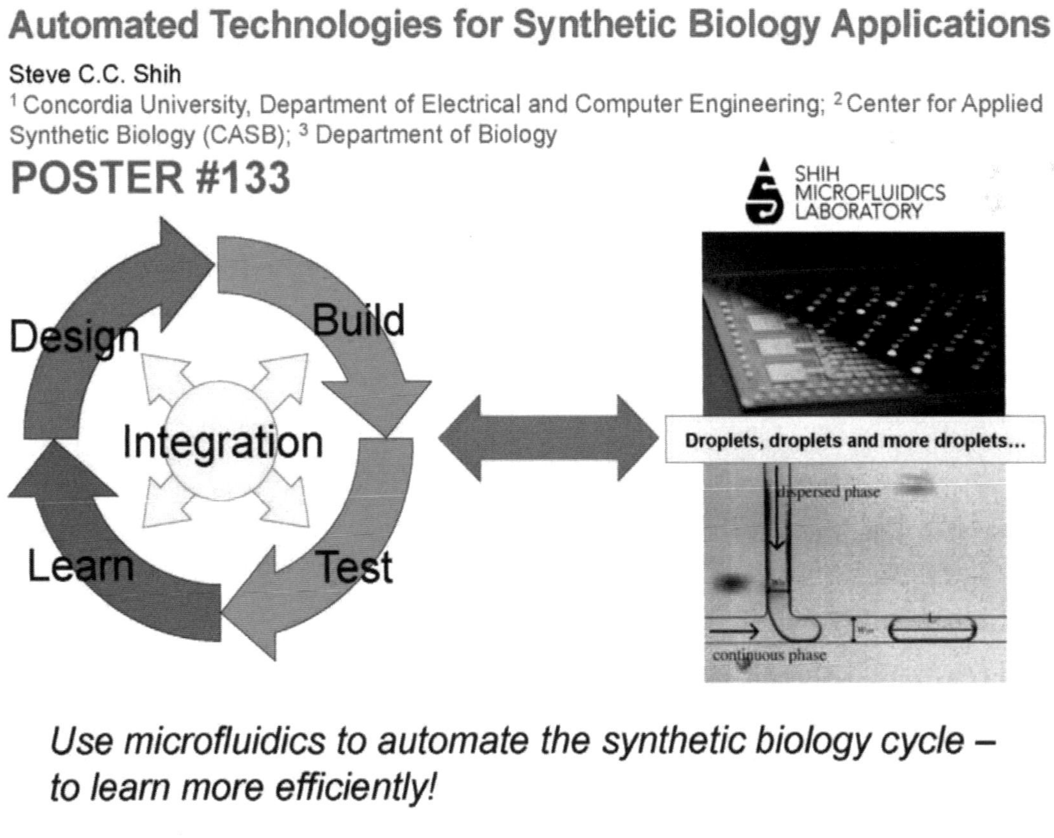

Automating the Design of Digital Microfluidic Devices for Synthetic Biology

- Conference: Synthetic Biology Engineering Evolution Design SEED
- Year: 2017
- Proceeding: 2017 Synthetic Biology: Engineering, Evolution & Design (SEED)
- Group: Poster Session
- Session:

  Confirmed Posters

Authors:
Alistar, M.
Baudisch, P.

Digital microfluidic devices manipulate and process fluids as individual droplets on an array of electrodes. The benefits of using such devices stem from microfluidics, as well as engineering. On the microfluidics side, the miniaturization of fluids leads to reduced reagent consumption and faster time-to-result. On the engineering side, the device is (a) *compact and portable* as the electrode array can be actuated directly off battery with no need for additional pressure or vacuum pumps; (b) *reconfigurable* as the electrodes can be used interchangeably and (c) *programmable* as the droplet movement can be programmed directly from the computer.

However, digital microfluidic devices have not been used yet in large scale in synthetic biology. Indeed, the use of a universal system (i.e., a single generic device that can be re-programed for each new bio-protocol) rises contamination issues, while a disposable-cartridge system comes with the difficulty of customized design.

We developed several algorithms to automate the design process of digital microfluidic cartridges. The main advantage of using automation over human design is trade-off analysis: the algorithms are better at balancing two or more objectives, while humans can optimize mainly for one, e.g., currently, researchers design new devices that reduce either the costs or the time-to-result, but have a hard time designing for both objectives.

Our tool bridges the design expertise requirements and supports the interaction between human editing and automatic design. Our algorithm optimizes for fabrication cost, reagent cost and time-to-result at the same time. Concretely, starting from a bio-protocol, our algorithm obtains a continuous solution space that allows selection among trade-off alternatives.

We tested our tool on seven bio-protocols commonly used in synthetic biology. The obtained designs were fabricated as real digital microfluidic cartridges and successfully tested for functionality.

By providing an automated way to efficiently generate digital microfluidic devices, we advance their use in synthetic biology research. New specific devices can be automatically generated for any new procedure that is established. Moreover, we believe that the use of our automated tool can expand beyond laboratory research, towards personal use.

Thus, our tool provides a step towards a future in which even non-technical users will be able to create microfluidic devices for their personal applications, thereby democratizing parts of health care. Our future work is focused on exploring the use of a digital microfluidic devices as personal laboratories operated at home.

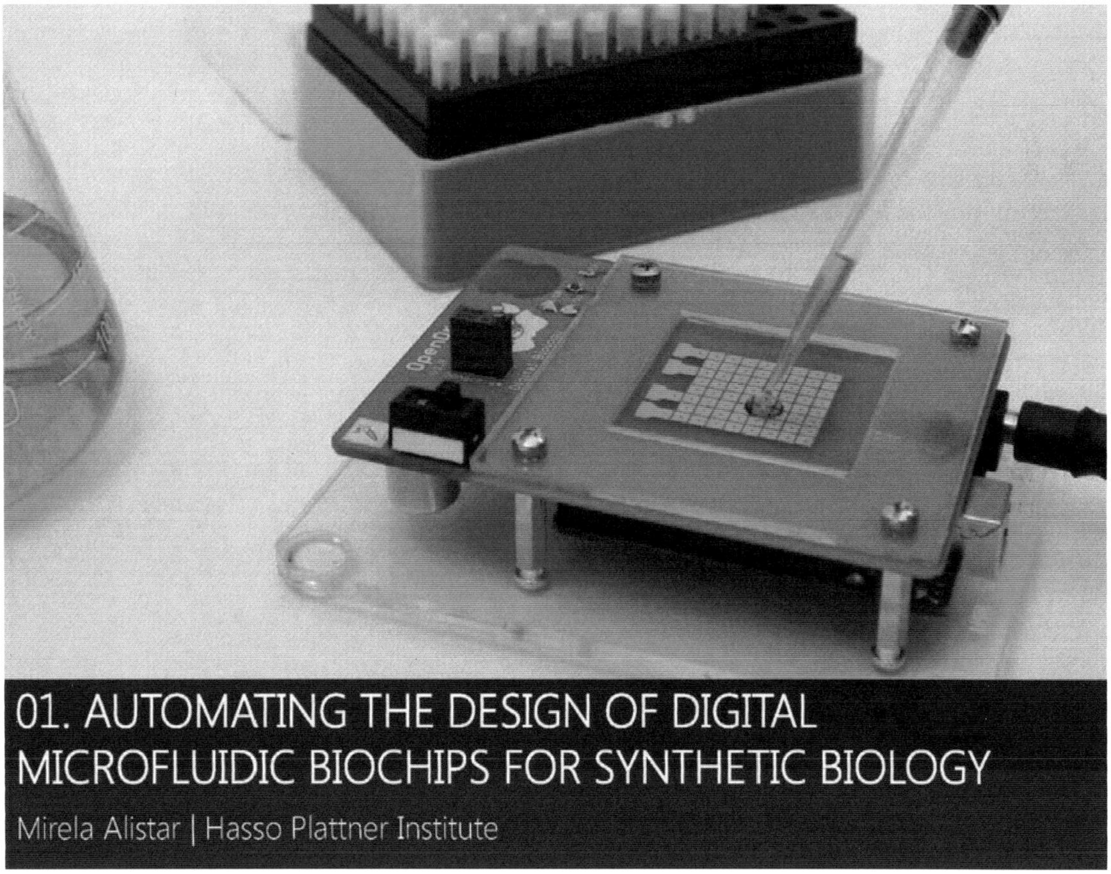

Bacterial Biosensors Based on Ligand-Dependent Stability

- Conference: Synthetic Biology Engineering Evolution Design SEED
- Year: 2017
- Proceeding: 2017 Synthetic Biology: Engineering, Evolution & Design (SEED)
- Group: Poster Session
- Session:

  Confirmed Posters

Authors:
Brandsen, B., University of Washington
Mattheisen, J., University of Washington
Fields, S., University of Washington
Our ability to predict protein expression from DNA sequence alone remains poor, reflecting our limited understanding of cis-regulatory grammar and hampering the design of engineered genes for synthetic biology applications. We have generated a model that predicts the translational efficiency of the 5â□™ untranslated region (UTR) of mRNAs in the yeast *Saccharomyces cerevisiae*. We constructed a library of nearly half a million 50 nucleotide-long random 5â□™ UTRs and assayed them in a single massively parallel growth selection experiment. The resulting data have allowed us to quantify the impact on translation of Kozak sequence composition, upstream open reading frames (uORFs) and secondary structure. With this data, we have trained a convolutional neural network on the random library and validated it by predicting the translational efficiency of 5â□™ UTRs that natively occur in yeast. The model additionally was used to computationally evolve highly translating 5â□™ UTRs. We have confirmed experimentally that the great majority of the evolved sequences lead to higher translation rates than the starting sequences, demonstrating the predictive utility of this model.

Biodevisor: Automated Refactoring of Biosynthetic Gene Clusters

- Conference: Synthetic Biology Engineering Evolution Design SEED
- Year: 2017
- Proceeding: 2017 Synthetic Biology: Engineering, Evolution & Design (SEED)
- Group: Poster Session
- Session:

  Confirmed Posters

Authors:
Oberortner, E., Lawrence Berkeley National Laboratory
Deutsch, S., Lawrence Berkeley National Laboratory
Hadjithomas, M., DOE Joint Genome Institute

Cell surface glycans make up an important mode of intercellular communication and the enzymes that assemble and degrade them are not only important targets for therapeutics, but also powerful tools for chemical glycobiology. Our recent work has explored 1) the use of engineered glycoside hydrolases for generating universal blood by antigen removal, 2) the identification of inhibitors to glycosyltransferases in cancer pathways, and 3) the development of engineered â□œglycosynthasesâ□□ to assemble and probe the function of glycosaminoglycan structures.

We have previously demonstrated that blood antigen-cleaving glycoside hydrolases can be engineered as tools towards the generation of universal blood through the cleavage of carbohydrate blood group antigens. Identification and directed evolution of these enzymes is enabled by a strategy that involves high-throughput screening with a sensitive fluorescence based assay. Using a similar approach, creatively repurposed to assay for glycosyltransferase rather than glycoside hydrolase activity, we have developed a high-throughput screen for inhibitors targeting fucosyltransferase enzymes that are highly expressed in cancers. Such inhibitors have potential as therapeutics for certain cancers wherein these enzymesâ□™ upregulated activity results in the assembly of carbohydrate structures (e.g. sialyl Lewis^{X}) involved in malignant processes including metastasis. In another focus, we are developing tools for chemoenzymatic glycan synthesis through mechanism-based engineering of â□œglycosynthaseâ□□ enzyme biocatalysts to assemble specific glycosaminoglycan structures. A facile enzymatic synthesis of these glycans will enable glycobiologists to probe their interactions in processes such as neural development and cancer.

Biosensor Expansion Out of an Ancient Evolutionary Cul-de-Sac

- Conference: Synthetic Biology Engineering Evolution Design SEED
- Year: 2017
- Proceeding: 2017 Synthetic Biology: Engineering, Evolution & Design (SEED)
- Group: Poster Session
- Session:

  Confirmed Posters

Authors:
Ellefson, J. W., University of Texas at Austin
Ellington, A. D., The University of Texas at Austin
Ledbetter, M. P.
Microbes can be engineered to produce biologically-derived replacements for gasoline, diesel, and jet fuel. However, a major technical challenge is that cells can only tolerate limited concentrations of typical fuels. Hydrocarbons are toxic because they damage the cell membrane and increase membrane fluidity. Thus, without further engineering there is a fundamental upper limit to how much biofuel microbes can produce, and cells must trade off production against survival. However, there are many natural examples of microbes that thrive in oil rich environments, such as around natural oil seeps or near oil spills. These microbes harbor mechanisms for tolerating the toxic effects of hydrocarbons. By using library-based approaches, we have identified several novel strategies for improving microbial biofuel tolerance by selecting for genes that enable survival. For example, expression of efflux pumps can dramatically improve tolerance in production strains, and we have further improved survival by designing synthetic feedback systems to control pump expression. This talk will discuss our efforts to enhance microbial synthesis of next-generation biofuels by developing tools for improving microbial tolerance of biofuel production conditions.

Bistable Switches for Yeast Population Growth Designed on Predictions from a Toolkit of Cell Division Regulators

- Conference: Synthetic Biology Engineering Evolution Design SEED
- Year: 2017
- Proceeding: 2017 Synthetic Biology: Engineering, Evolution & Design (SEED)
- Group: Poster Session
- Session:

  Confirmed Posters

Authors:
Carignano, A., University of Washington
Adam, L.
de Lange, O.
Klavins, E., University of Washington

Positive feedback loop gene circuits are thought to control the process of cell differentiation. One such example is the cell-cycle lengthening mechanism that, through protein accumulation, leads to myeloid differentiation in stem cells [Kueh et al, 2013] and, the differentiation of neural precursors [Salomoni et al, 2010]. In both systems, a regulatory factor slows down cell division leading to its own accumulation, which constitutes de facto a positive feedback between the cell cycle and the factor itself. This naturally occurring principle has been exploited to construct a synthetic circuit that causes bistable gene expression in E. coli [Tan et al, 2009]. However, we propose to modulate the growth rate to design synthetic switches.

Here, we propose a library of parts to control/tune cell growth in yeast, and we show how we used it to rationally design a growth-based bistable switch. Through a literature search, we identified Yeast genes susceptible to induce growth arrest and use their antagonists to modulate the growth rate. Each gene was then expressed from an inducible promoter and resultant growth rates quantified. Finally, we used data-trained mathematical models to have a quantitative and qualitative description of each gene and their effects on cell division. We made use of this mathematical characterization to search for combinations of parts that leads to growth bistability. In a first attempt to build a growth-based bistable switch, we found that not only were the yeast cells were able to switch between the two states at any given time but also, the bistable switch remained in either state for over 5 days without loss of function.

In summary, we constructed and characterized a vocabulary of parts to control the growth rate of Yeast cells.Â We mathematically demonstrated that these parts could be used to systematically construct several growth-dependent behaviors in a predictable fashion. As a proof of concept, we designed and built the first stable growth-based bistable switch in yeast, and described a method that could be easily used to increase the number of these systems.

Bistable switches for yeast population growth designed on predictions from a toolkit of cell division regulators

Alberto Carignano, Laura Adams, Orlando de Lange, and Eric Klavins

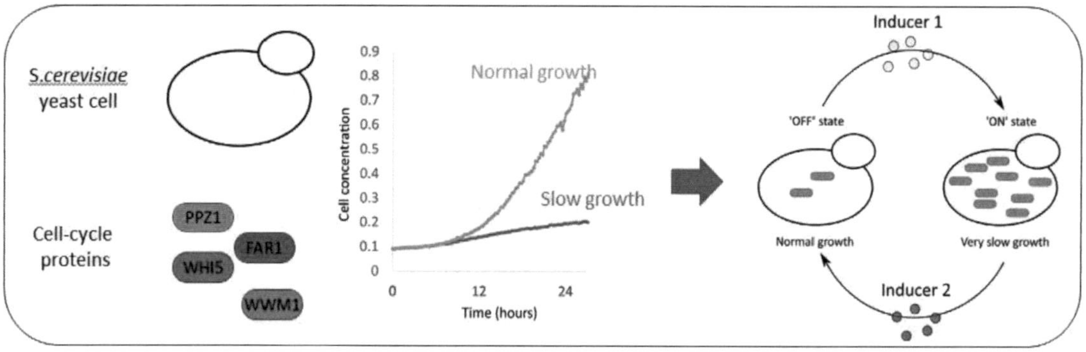

We are using this system in the area of:

- **Homeostasis control**
- **Cell communication**
- **Cell differentiation**

POSTER #14

Don't miss this one!

Cas9 Activity in Human Chromatin: Inhibition and Strategies for Increased Efficiency

- Conference: Synthetic Biology Engineering Evolution Design SEED
- Year: 2017
- Proceeding: 2017 Synthetic Biology: Engineering, Evolution & Design (SEED)
- Group: Poster Session
- Session:

  Confirmed Posters

Authors:
Daer, R., Arizona State University
Haynes, K., Arizona State University
Barrett, C., Arizona State University

The prokaryotic CRISPR/Cas9 system enables precise changes at genomic targets. Due to the ease of its customization, Cas9 promises to make high-throughput genetic modifications possible. However, as Cas9 is a bacterially derived system, it has not evolved to interact with the mammalian genome which is compartmentalized and packaged into a nucleo-protein complex known as chromatin. We investigated the mechanism by which dynamic human chromatin states impede efficient Cas9 editing and identified a potential strategy to improve Cas9 efficiency. Using a transgenic HEK293 cell line, we induced accumulation of a Polycomb-based repressor (Gal4-EED) at a luciferase reporter gene to achieve partially and fully silenced chromatin states. We used a transactivator (Gal4-p65) to stimulate a hyperactive state. At silenced luciferase, we confirmed accumulation of the silencing mark H3K27me3 along luciferase by ChIP-qPCR. We measured Cas9-mediated editing using SURVEYOR assay and Cas9 binding using ChIP-qPCR. Interestingly, of the nine gRNA target sites tested along luciferase, Cas9 binding and cutting were inhibited at just five of the sites; four of the target sites were edited at a similar efficiency in each chromatin state. We also observed that a hyperactive state driven by Gal4-p65 reduces Cas9 editing, suggesting high expression levels could be a barrier to efficient Cas9 targeting. To determine a strategy for improving CRISPR efficiency, we broadly inhibited silenced chromatin with either siRNA against Polycomb or the chromatin disrupting drug UNC1999. Increased luciferase expression, indicating an active open chromatin state, was accompanied by increased Cas9 editing efficiency. Finally, we have designed a new transgenic system (inducible Gal4-chromatin remodeler fusions) to determine whether a DNA-targeted chromatin-remodeling proteins can open chromatin at a single locus for more precise enhancement of CRISPR accessibility. Ongoing work includes mapping of nucleosomes at CRISPR target sites to gain a deeper understanding of Cas9-chromatin interaction in human cells. So far, our work represents a significant advance towards understanding the mechanism and extent of Cas9-inhibition by dynamic chromatin. Artificial chromatin remodeling offers a promising solution for efficient binding and editing at sites located in silenced chromatin.

Cas9 activity in human chromatin: Inhibition and strategies for increased efficiency

Poster 26

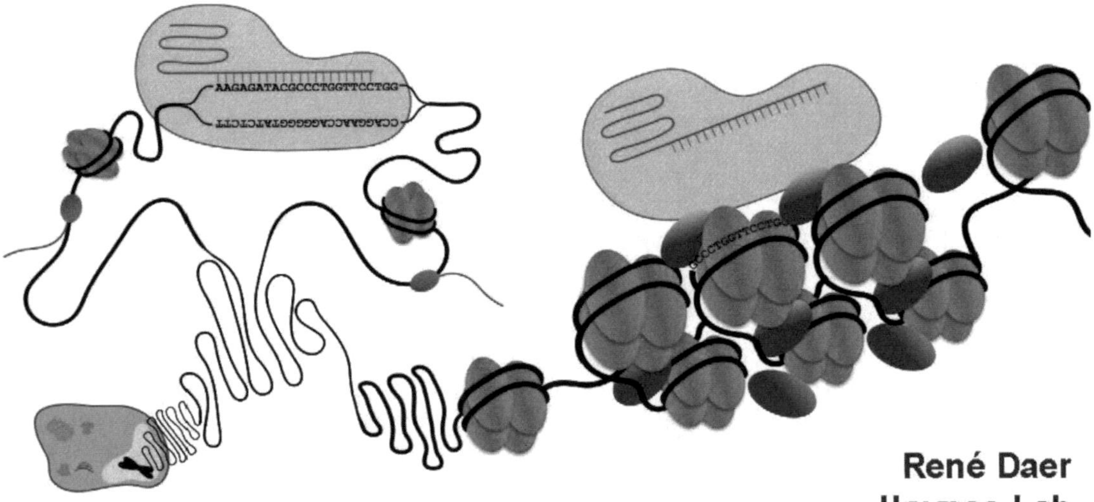

René Daer
Haynes Lab
Arizona State University

Cell-Sized Mechanosensitive and Biosensing Compartment Programmed with DNA

- Conference: Synthetic Biology Engineering Evolution Design SEED
- Year: 2017
- Proceeding: 2017 Synthetic Biology: Engineering, Evolution & Design (SEED)
- Group: Poster Session
- Session:

  Confirmed Posters

Authors:
Majumder, S.
Liu, A. P., University of Michigan
Noireaux, V., University of Minnesota
Garamella, J., University of Minnesota
Wang, Y. L.
Our ability to design cell factories to produce valuable chemicals requires the â□œrecombinationâ□□ not only of existing but also designer enzymes into novel metabolic pathways to achieve entirely new metabolic function. This poses two distinct challenges that need to be solved in an integrated way if we ambition to fully deliver on the promise of synthetic biology. The first challenge deals with the design, at scale, of novel enzymes with high level of activities for reactions not known to be catalyzed in nature. To this end, Arzeda has developed and scaled high-throughput computational methodologies to exploit natural enzyme latent catalytic promiscuity and rapidly design new biocatalysts. A successful industrial application of this technology will be presented.

The second challenge is that of finding optimal ways to arrange natural and designed enzymes to biosynthetize a small-molecule of interest. To illustrate our progress towards solving this challenge, we will discuss a novel computational tool that Arzeda has developed for the automated design of novel biosynthetic pathways. Inspired by retrosynthetic methods in organic chemistry, our software draws on databases of known natural enzymatic reactions as well as reactions that can be catalyzed by designed enzymes to exhaustively enumerate biosynthetic routes from a set of desired metabolites down to any small molecule of interest. Pathways are ranked based on thermodynamic feasibility and other metrics. We will discuss applications of industrial relevance in the field of fine and bulk chemicals.

Characterizing the Role of 23S rRNA Peptidyltransferase Center Nucleotides in Translation

- Conference: Synthetic Biology Engineering Evolution Design SEED
- Year: 2017
- Proceeding: 2017 Synthetic Biology: Engineering, Evolution & Design (SEED)
- Group: Poster Session
- Session:

  Confirmed Posters

Authors:
d'Aquino, A., Northwestern University
Jewett, M. C., Northwestern University
Azim, T., Northwestern University
Amyris is an end-to-end biological manufacturer that uses synthetic biology to engineer microbes as catalysts for conversion of sugar into chemical products. Many of these chemical targets were previously derived from limited natural resources and economically inaccessible via traditional chemistries. In the last decade, Amyris has developed a sophisticated technology platform for engineering microbes to produce high levels of isoprenoids including artemisinin and farnesene. In recent years, we have begun to harness the power of this proven infrastructure to expand the portfolio of bio-manufactured chemical products.

In this talk, I will describe the current state-of-the-art tools at Amyris, including our automated strain engineering, high-throughput screening, and fermentation. Then I will discuss how these tools are evolving under the DARPA-funded â□œMgs to Kgsâ□□ project to enable the concurrent development of 100â□™s of products simultaneously. In the first year, the project has produced more than 150 target molecules from all parts of metabolic space. As the project continues, â□œMgs to Kgsâ□□ will enable Amyris to incorporate next generation high-throughput â□˜omics, automation, data science, and machine learning tools to further increase the rate at which biology can be engineered.

Characterizing Nopaline Synthase Promoter Variants for Use in Plant Synthetic Biology

- Conference: Synthetic Biology Engineering Evolution Design SEED
- Year: 2017
- Proceeding: 2017 Synthetic Biology: Engineering, Evolution & Design (SEED)
- Group: Poster Session
- Session:

  Confirmed Posters

Authors:
Morey, K., Colorado State University
Medford, J. I., Colorado State University

Digital microfluidic devices manipulate and process fluids as individual droplets on an array of electrodes. The benefits of using such devices stem from microfluidics, as well as engineering. On the microfluidics side, the miniaturization of fluids leads to reduced reagent consumption and faster time-to-result. On the engineering side, the device is (a) *compact and portable* as the electrode array can be actuated directly off battery with no need for additional pressure or vacuum pumps; (b) *reconfigurable* as the electrodes can be used interchangeably and (c) *programmable* as the droplet movement can be programmed directly from the computer.

However, digital microfluidic devices have not been used yet in large scale in synthetic biology. Indeed, the use of a universal system (i.e., a single generic device that can be re-programed for each new bio-protocol) rises contamination issues, while a disposable-cartridge system comes with the difficulty of customized design.

We developed several algorithms to automate the design process of digital microfluidic cartridges. The main advantage of using automation over human design is trade-off analysis: the algorithms are better at balancing two or more objectives, while humans can optimize mainly for one, e.g., currently, researchers design new devices that reduce either the costs or the time-to-result, but have a hard time designing for both objectives.

Our tool bridges the design expertise requirements and supports the interaction between human editing and automatic design. Our algorithm optimizes for fabrication cost, reagent cost and time-to-result at the same time. Concretely, starting from a bio-protocol, our algorithm obtains a continuous solution space that allows selection among trade-off alternatives.

We tested our tool on seven bio-protocols commonly used in synthetic biology. The obtained designs were fabricated as real digital microfluidic cartridges and successfully tested for functionality.

By providing an automated way to efficiently generate digital microfluidic devices, we advance their use in synthetic biology research. New specific devices can be automatically generated for any new procedure that is established. Moreover, we believe that the use of our automated tool can expand beyond laboratory research, towards personal use.

Thus, our tool provides a step towards a future in which even non-technical users will be able to create microfluidic devices for their personal applications, thereby democratizing parts of health care. Our future work is focused on exploring the use of a digital microfluidic devices as personal laboratories operated at home.

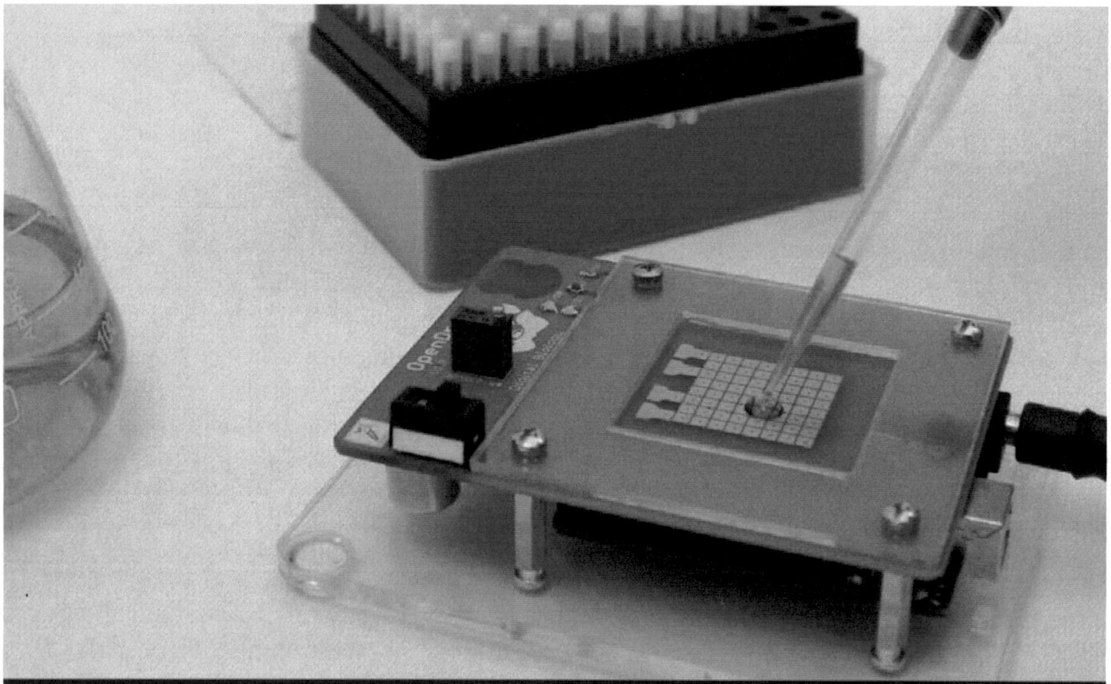

01. AUTOMATING THE DESIGN OF DIGITAL MICROFLUIDIC BIOCHIPS FOR SYNTHETIC BIOLOGY

Mirela Alistar | Hasso Plattner Institute

Cloning, Expression and Characterization of a Self-Sufficient Cytochrome P450 Monooxygenase from Bacillus Species

- Conference: Synthetic Biology Engineering Evolution Design SEED
- Year: 2017
- Proceeding: 2017 Synthetic Biology: Engineering, Evolution & Design (SEED)
- Group: Poster Session
- Session:

  Confirmed Posters

Authors:
Rimal, H.
Dangi, B.
Oh, T. J.
Plants are a rich source of unique scaffolds, including 25% of natural-product-derived drugs. However, the discovery, synthesis, and overall material supply chains for sourcing plant natural products and their derivatives remain ad hoc, biased, and tedious. While microbial biosynthesis presents compelling alternatives to traditional approaches based on extraction from natural plant hosts, many challenges exist in the reconstruction of plant specialized metabolic pathways in microbial hosts. My laboratory has developed approaches to address the challenges that arise in the reconstruction of complex biosynthesis schemes, including spatial engineering strategies to direct the activities and specificities of pathway enzymes and recoding strategies to address folding, processing, and stability issues that may arise with the expression of plant enzymes in heterologous microbial hosts. We have utilized these strategies to develop yeast-based production platforms for an important class of plant alkaloids, the benzylisoquinoline alkaloids, including the medicinal opioids. These synthetic biology platforms will lead to transformative advances in natural product discovery, drug development, and production.

Co-Production of Platform Chemicals 1, 3 Propane Diol and 3-Hydroxy Propionic Acid from Waste Glycerol By Clostridium Pasteurianum

- Conference: Synthetic Biology Engineering Evolution Design SEED
- Year: 2017
- Proceeding: 2017 Synthetic Biology: Engineering, Evolution & Design (SEED)
- Group: Poster Session
- Session:

  Confirmed Posters

Authors:
Ortega, D., University of Nottingham
Schuchmann, K., University of Nottingham
Minton, N., The University of Nottingham
Zhang, Y., University of Nottingham

Clostridium are a family of obligate anaerobic and spore forming bacteria found in diverse environments with roles in clinical pathogenesis but also known for their fermentative abilities. In the early 20th century, fermentation of starch into acetone, butanol and ethanol (later known as ABE fermentation) by *Clostridium acetobutylicumÂ* was one of the first commercial applications of a microbial biocatalyst for production of industrial chemicals.

Interest in *Clostridium pasteurianum* has been reignited in recent years as it is one of the few species of *ClostridiumÂ* which is able to ferment glycerol â□" a waste by-product of the biodiesel industry â□" into industrially relevant chemicals n-butanol and 1,3-propanediol (1,3 PDO). 3-Hydroxypropionic acid (3HP) is another industrially relevant chemical for which biological production routes have been pursuedÂ as it can easily be dehydrated into acrylic acid. Acrylic acid and 1,3 PDO have forecasted global market values in excess of $13 and $0.5 billion by 2020 respectively.

Using synthetic biology and rational metabolic engineering approaches, we describe the genetic manipulation of *C. pasteurianum* for the co-production of 1,3 PDO and 3HP from waste glycerol. Aspects of glycerol metabolism have been investigated towards gaining a better understanding of variable gene expression of *C. pasteurianum* during glycerol fermentation. Heterologous expression of enzymes required for production of 3HP are introduced. Both Co-A dependent and Co-A independent pathways for 3HP production from glycerol are explored. Genes involved in production of unwanted by-products are removed towards enhancing product yield. Effects of genetic modifications on metabolite equilibriums and redox balance. Strain engineering strategies by CRISPR and allelic exchange are employed. Generic lessons learned with broad range applicability for production of other bio commodities using different microbial platforms are discussed.

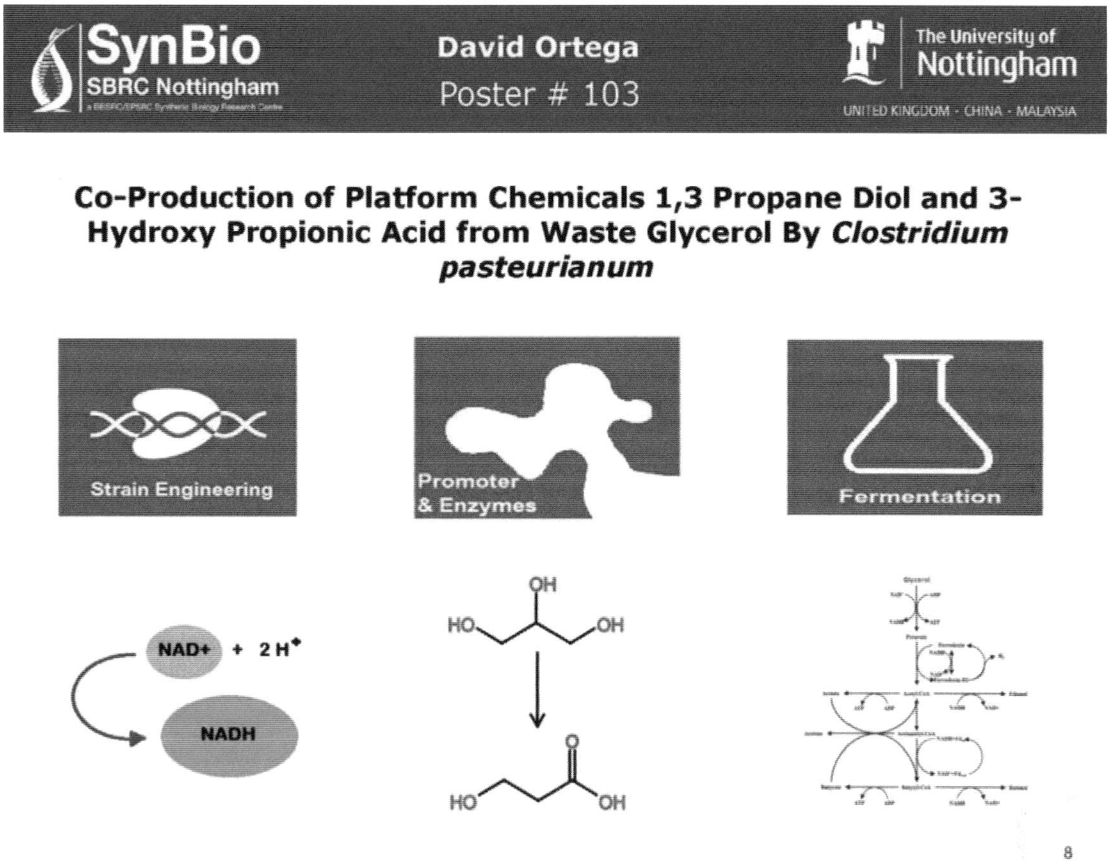

Composition and Prediction of Multi-Input Promoters

- Conference: Synthetic Biology Engineering Evolution Design SEED
- Year: 2017
- Proceeding: 2017 Synthetic Biology: Engineering, Evolution & Design (SEED)
- Group: Poster Session
- Session:

  Confirmed Posters

Authors:
Zong, D., Rice University
Bennett, M. R., Rice University
Ott, W., University of Houston
Gupta, C., University of Houston

Synthetic Biology aims to implement new cellular features through the design of original genetic parts and systems. A wide variety of circuit architectures have been built successfully, proving the ability of synthetic genetic parts to predictably interact with each other within a single bacterial cell [4]. The next stage of synthetic biology will involve the *de novo* engineering of self-organizing multicellular systems. One of the main methods found in nature for creating complex spatially distributed systems is cellular differentiation through asymmetric cell division (ACD), whereby stem cells spawn off irreversibly differentiated daughters. The presence of different cell lineages within an eukaryotic tissue allows for the partitioning of metabolic and structural tasks. Furthermore, if one or both of the daughter cells becomes incapable of dividing, controlling asymmetric division probability can lead to tissue size and shape control [5]. In complex eukaryotic organisms ACD is typically brought about by the differential expression of key genes within each cell type. Here, we engineered a novel method for generating ACD in *Escherichia coli* and regulating its frequency within the population. This work provides a platform for the construction of complex synthetic tissues and organisms from bacterial cells that differentiate during development of the population.

Current methods for creating differentiated cell types in synthetic bacteria utilize gene regulatory networks that allow for bistable gene expression [2, 1]. While such systems generate bimodal populations, they are very sensitive to both intrinsic and extrinsic noise sources that can randomly force transitions between states [3]. We use a novel method for creating terminally differentiated synthetic bacterial cells. Specifically, we refactored DNA shuttling mechanisms from *Caulobacter crescentus*

into *E. coli* to target plasmid DNA into one and only one daughter cell upon division. Thus, the two daughter cells will not just have different gene expression states, but will be irreversibly genetically distinct. Our findings will lead to synthetic "tissues" that can be used in complex settings.

References

[1] R. G. Egbert and E. Klavins. Fine-tuning gene networks using simple sequence repeats.*Proceedings of the National Academy of Sciences*, 109(42):16817â□“16822, 2012.
[2] T. S. Gardner, C. R. Cantor, and J. J. Collins. Construction of a genetic toggle switch in *Escherichia coli*. *Nature*, 403(6767):339â□“342, 2000.
[3] C. Gupta, J. M. LÃ³pez, W. Ott, K. JosiÄ‡, and M. R. Bennett. Transcriptional delay stabilizes bistable gene networks. *Physical review letters*, 111(5):058104, 2013.
[4] A. S. Khalil and J. J. Collins. Synthetic biology: applications come of age. *Nature Reviews*

Genetics, 11(5):367â□“379, 2010.
[5] A. D. Lander, K. K. Gokoffski, F. Y. Wan, Q. Nie, and A. L. Calof. Cell lineages and the logic of proliferative control. *PLoS Biol*, 7(1):e1000015, 2009.

POSTER # 93 - SARA MOLINARI - RICE UNIVERSITY The Biodesign Lab RICE

ENGINEERED CONTROL SYSTEMS FOR ASYMMETRIC CELL DIVISION IN SYNTHETIC BACTERIA

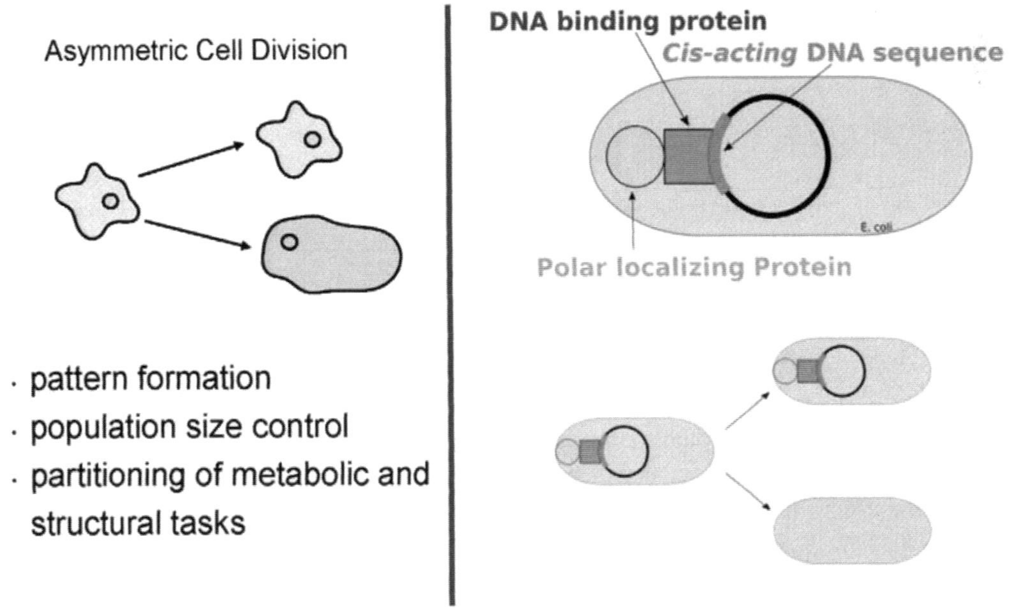

Computational Design of Self-Assembling Protein Nanomachines

- Conference: Synthetic Biology Engineering Evolution Design SEED
- Year: 2017
- Proceeding: 2017 Synthetic Biology: Engineering, Evolution & Design (SEED)
- Group: Poster Session
- Session:

  Confirmed Posters

Authors:
Courbet, A., University of Washington
Wei, K., University of Washington
Nattermann, U., University of Washington
Moyer, A., University of Washington
Hsia, Y., University of Washington
Ueda, G., University of Washington
Fallas, J., University of Washington
Boyken, S., University of Washington
Chen, Z., University of Washington
Xu, C., University of Washington
Ceze, L.
Daniel, T.
Smith, J.
Bradley, P.
Baker, D., University of Washington
Ben-Sasson, A., U
Vulovic, I., University of Washington
CRISPR/Cas9 has wide-reaching scientific and therapeutic applications, but efficient, exact targeting of cutting sites can be challenging. Several cas9 variants have been engineered to target alternate PAMs; however, generation of new variants is often laborious. Additionally, the cutting ability of cas9 nucleases is often highly guide-dependent and shows a large amount of variation between sites. In order to engineer new nuclease variants targeted to cleave a given site, we developed a directed evolution platform to select for DNA cleavage in bacteria. Our system allows rapid generation of mutant libraries, which are challenged with phage containing target sites in a competitive pool. Using this method, we developed *S. aureus* cas9 variants which have altered PAM preferences to target cleavage to novel targets of therapeutic relevance.

Computational Design of Synthetic Co-Transcriptional RNA Elements for the Dynamic Regulation of Gene Expression

- Conference: Synthetic Biology Engineering Evolution Design SEED
- Year: 2017
- Proceeding: 2017 Synthetic Biology: Engineering, Evolution & Design (SEED)
- Group: Poster Session
- Session:

  Confirmed Posters

Authors:
Sparkman-Yager, D. W.
Recent studies into the interplay between the microbiome and host have extensively implicated short-chain fatty acid (SCFA) fermentation products as contributors to both intra and extra intestinal health. In the context of the gut, butyrate (a four carbon SCFA) serves as the primary energy source for epithelial colonocytes, drives regulatory T cell differentiation and is therefore essential for the maintenance of colonic mucosal health. On a systemic level, butyrate may influence insulin sensitivity, cholesterol synthesis and autoimmunity. Typically, butyrate is produced by the action of slow-growing Gram positive organisms â“ often Clostridial in origin â“ during the fermentation of cellulose and complex carbohydrates. To counteract the depletion of butyrate in the gut, we iteratively engineered *Escherichia coli* Nissle 1917 to produce butyrate. The new butyrate-producing Nissle was capable of millimolar level production of butyrate comparable with the levels produced by enteric Clostridial species. The circuitry was placed under anaerobic control to support induction upon entry into the gastrointestinal track. Finally, proof-of-concept *in vivo* experimentation showed production of butyrate in the mouse gut. This engineered strain may be useful for the treatment of a variety of gastrointestinal and systemic conditions associated with SCFA deficiency.

Controlling Bdellovibrio Bacteriovorus Gene Expression and Predation Using Synthetic Riboswitches

- Conference: Synthetic Biology Engineering Evolution Design SEED
- Year: 2017
- Proceeding: 2017 Synthetic Biology: Engineering, Evolution & Design (SEED)
- Group: Poster Session
- Session:

  Confirmed Posters

Authors:
Dwidar, M.
Yokobayashi, Y.

Improving the RNA switch design process through big data and machine learning

Calvin M. Schmidt, Christina D. Smolke
Stanford University; Stanford, California

RNA switches are a class of functional RNAs that support conditional regulation of gene expression. These genetic switches have broad application in the design of biological systems responsive to changes in their environment with implications for the fields of metabolic engineering, environmental sensing, and therapeutics. The mechanisms underlying the activity of RNA switches rely on programmed structural conformational changes encoded in the sequence of the RNA molecule, implying that an improved elucidation of relationships between sequence, structure, and function can be used to improve our capacities for de novo design. One type of RNA switch is the ribozyme switch, in which binding of the target ligand changes the structure and thus cleavage activity of the switch. Here, we leverage sequencing data and machine learning methods to relate the sequence and structure of RNA molecules to their function and develop models that predict the function of novel ribozyme switches. Through prior work, we have access to data on the activity of hundreds of thousands of ribozyme switch sequences. Using automated structural analysis and machine learning methods, we leveraged a wealth of sequence-activity data on hundreds of thousands of ribozyme switches to develop mathematical models that predict the in vivo activity of a given ribozyme based on its sequence. We have used these models for the de novo design of novel ribozyme switches that exhibit changes in in vivo gene-regulatory activity upon introduction of a target ligand. Our work demonstrates that computational tools can be used to convert RNA sequences into quantitative features that represent key sequence and structural motifs and that these features can be used to develop models that can learn how interactions between these motifs translates to the activity of the functional RNA molecule. The predictive models we developed provide a broadly accessible tool for designing custom biosensors for use in engineering novel biological systems

Rational design of tertiary ribozyme switches through big data and machine learning

Calvin Schmidt
Stanford University
Poster 128

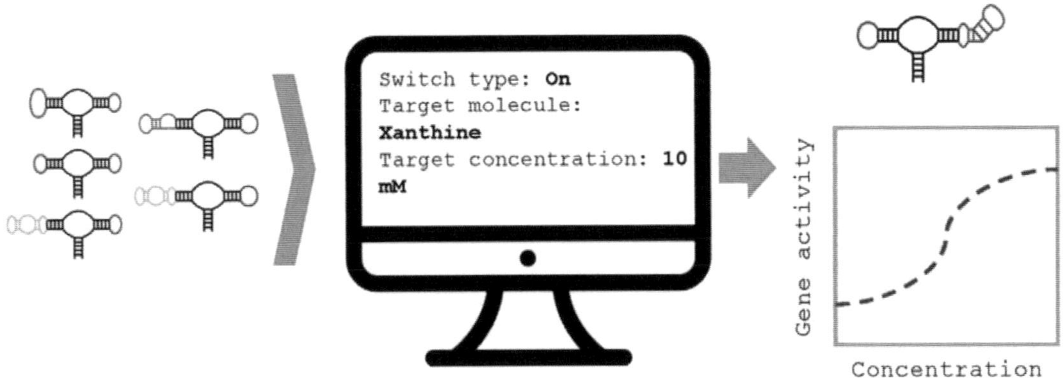

Controlling Energy Flow in Bacteria Using Engineered Ligand-Responsive Protein Electron Carriers

- Conference: Synthetic Biology Engineering Evolution Design SEED
- Year: 2017
- Proceeding: 2017 Synthetic Biology: Engineering, Evolution & Design (SEED)
- Group: Poster Session
- Session:

  Confirmed Posters

Authors:
Atkinson, J. T., Rice University
Silberg, J. J., Rice University
Campbell, I. J., Rice University
Bennett, G. N., Rice University
Torres, J., Rice University

Synthetic Biology aims to implement new cellular features through the design of original genetic parts and systems. A wide variety of circuit architectures have been built successfully, proving the ability of synthetic genetic parts to predictably interact with each other within a single bacterial cell [4]. The next stage of synthetic biology will involve the *de novo* engineering of self-organizing multicellular systems. One of the main methods found in nature for creating complex spatially distributed systems is cellular differentiation through asymmetric cell division (ACD), whereby stem cells spawn off irreversibly differentiated daughters. The presence of different cell lineages within an eukaryotic tissue allows for the partitioning of metabolic and structural tasks. Furthermore, if one or both of the daughter cells becomes incapable of dividing, controlling asymmetric division probability can lead to tissue size and shape control [5]. In complex eukaryotic organisms ACD is typically brought about by the differential expression of key genes within each cell type. Here, we engineered a novel method for generating ACD in *Escherichia coli* and regulating its frequency within the population. This work provides a platform for the construction of complex synthetic tissues and organisms from bacterial cells that differentiate during development of the population.

Current methods for creating differentiated cell types in synthetic bacteria utilize gene regulatory networks that allow for bistable gene expression [2, 1]. While such systems generate bimodal populations, they are very sensitive to both intrinsic and extrinsic noise sources that can randomly force transitions between states [3]. We use a novel method for creating terminally differentiated synthetic bacterial cells. Specifically, we refactored DNA shuttling mechanisms from *Caulobacter crescentus*

into *E. coli* to target plasmid DNA into one and only one daughter cell upon division. Thus, the two daughter cells will not just have different gene expression states, but will be irreversibly genetically distinct. Our findings will lead to synthetic "tissues" that can be used in complex settings.

References

[1] R. G. Egbert and E. Klavins. Fine-tuning gene networks using simple sequence repeats.*Proceedings of the National Academy of Sciences*, 109(42):16817â□“16822, 2012.
[2] T. S. Gardner, C. R. Cantor, and J. J. Collins. Construction of a genetic toggle switch in *Escherichia coli*. *Nature*, 403(6767):339â□“342, 2000.
[3] C. Gupta, J. M. LÃ³pez, W. Ott, K. JosiÄ‡, and M. R. Bennett. Transcriptional delay stabilizes bistable gene networks. *Physical review letters*, 111(5):058104, 2013.
[4] A. S. Khalil and J. J. Collins. Synthetic biology: applications come of age. *Nature Reviews*

Genetics, 11(5):367â□“379, 2010.
[5] A. D. Lander, K. K. Gokoffski, F. Y. Wan, Q. Nie, and A. L. Calof. Cell lineages and the logic of proliferative control. *PLoS Biol*, 7(1):e1000015, 2009.

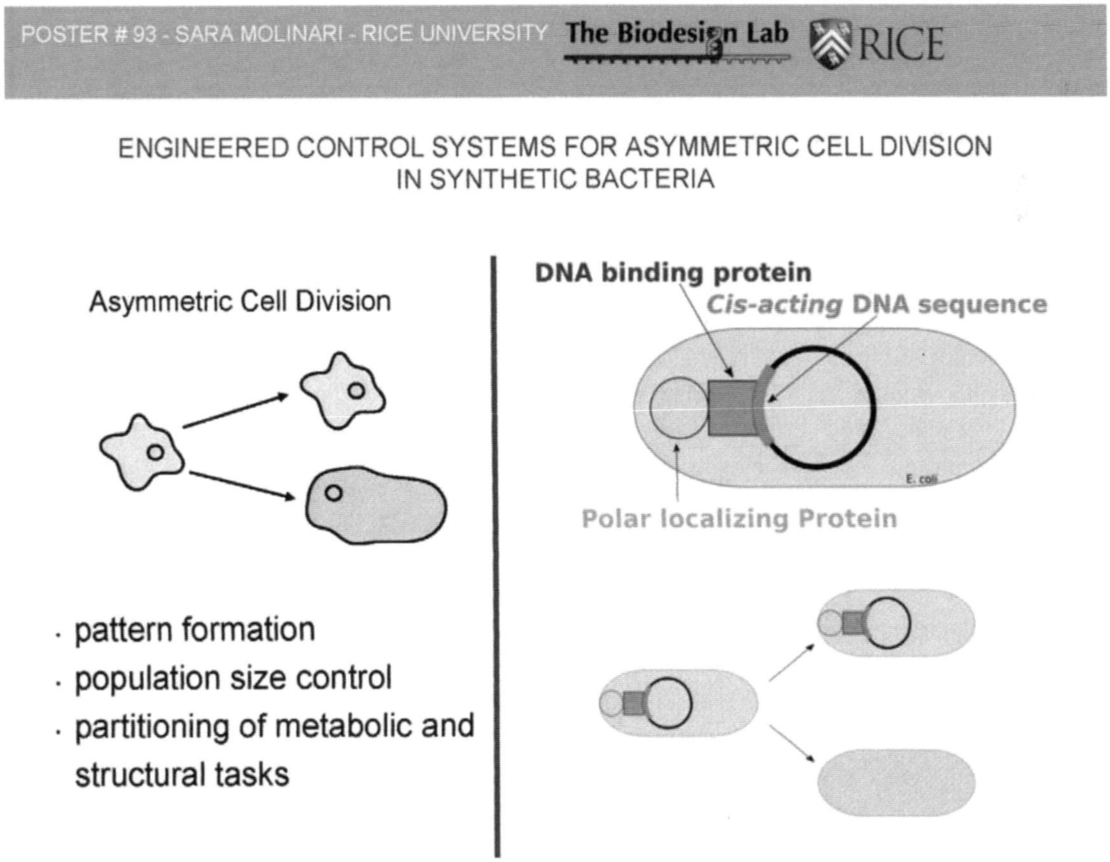

Converting Rhodosporidium Toruloides As a New Industrial Workhorse for the Production of Terpenes and Unusual Fatty Acids

- Conference: Synthetic Biology Engineering Evolution Design SEED
- Year: 2017
- Proceeding: 2017 Synthetic Biology: Engineering, Evolution & Design (SEED)
- Group: Poster Session
- Session:

  Confirmed Posters

Authors:
Ji, L.
Liu, Y., Temasek Life Sciences Laboratory
Our ability to predict protein expression from DNA sequence alone remains poor, reflecting our limited understanding of cis-regulatory grammar and hampering the design of engineered genes for synthetic biology applications. We have generated a model that predicts the translational efficiency of the 5â□™ untranslated region (UTR) of mRNAs in the yeast *Saccharomyces cerevisiae*. We constructed a library of nearly half a million 50 nucleotide-long random 5â□™ UTRs and assayed them in a single massively parallel growth selection experiment. The resulting data have allowed us to quantify the impact on translation of Kozak sequence composition, upstream open reading frames (uORFs) and secondary structure. With this data, we have trained a convolutional neural network on the random library and validated it by predicting the translational efficiency of 5â□™ UTRs that natively occur in yeast. The model additionally was used to computationally evolve highly translating 5â□™ UTRs. We have confirmed experimentally that the great majority of the evolved sequences lead to higher translation rates than the starting sequences, demonstrating the predictive utility of this model.

Coral: Python Abstractions for Genetic Design

- Conference: Synthetic Biology Engineering Evolution Design SEED
- Year: 2017
- Proceeding: 2017 Synthetic Biology: Engineering, Evolution & Design (SEED)
- Group: Poster Session
- Session:

  Confirmed Posters

Authors:
Bolten, N., University of Washington
Recent studies into the interplay between the microbiome and host have extensively implicated short-chain fatty acid (SCFA) fermentation products as contributors to both intra and extra intestinal health. In the context of the gut, butyrate (a four carbon SCFA) serves as the primary energy source for epithelial colonocytes, drives regulatory T cell differentiation and is therefore essential for the maintenance of colonic mucosal health. On a systemic level, butyrate may influence insulin sensitivity, cholesterol synthesis and autoimmunity. Typically, butyrate is produced by the action of slow-growing Gram positive organisms â“ often Clostridial in origin â“ during the fermentation of cellulose and complex carbohydrates. To counteract the depletion of butyrate in the gut, we iteratively engineered *Escherichia coli* Nissle 1917 to produce butyrate. The new butyrate-producing Nissle was capable of millimolar level production of butyrate comparable with the levels produced by enteric Clostridial species. The circuitry was placed under anaerobic control to support induction upon entry into the gastrointestinal track. Finally, proof-of-concept *in vivo* experimentation showed production of butyrate in the mouse gut. This engineered strain may be useful for the treatment of a variety of gastrointestinal and systemic conditions associated with SCFA deficiency.

Creating Single-Copy Genetic Circuits and Microbial Kill Switches

- Conference: Synthetic Biology Engineering Evolution Design SEED
- Year: 2017
- Proceeding: 2017 Synthetic Biology: Engineering, Evolution & Design (SEED)
- Group: Poster Session
- Session:

  Confirmed Posters

Authors:
Lee, J. W., Harvard University
Collins, J. J., Massachusetts Institute of Technology
Our ability to predict protein expression from DNA sequence alone remains poor, reflecting our limited understanding of cis-regulatory grammar and hampering the design of engineered genes for synthetic biology applications. We have generated a model that predicts the translational efficiency of the 5â□™ untranslated region (UTR) of mRNAs in the yeast *Saccharomyces cerevisiae.* We constructed a library of nearly half a million 50 nucleotide-long random 5â□™ UTRs and assayed them in a single massively parallel growth selection experiment. The resulting data have allowed us to quantify the impact on translation of Kozak sequence composition, upstream open reading frames (uORFs) and secondary structure. With this data, we have trained a convolutional neural network on the random library and validated it by predicting the translational efficiency of 5â□™ UTRs that natively occur in yeast. The model additionally was used to computationally evolve highly translating 5â□™ UTRs. We have confirmed experimentally that the great majority of the evolved sequences lead to higher translation rates than the starting sequences, demonstrating the predictive utility of this model.

CRISPR Enabled Trackable Genome Engineering for Isopropanol Production in Escherichia coli

- Conference: Synthetic Biology Engineering Evolution Design SEED
- Year: 2017
- Proceeding: 2017 Synthetic Biology: Engineering, Evolution & Design (SEED)
- Group: Poster Session
- Session:

  Confirmed Posters

Authors:
Liang, L., University of Colorado Boulder
Liu, R., University of Colorado Boulder
Garst, A., Muse Biotechnology Inc.
Lee, T.
Nogué, V. S. I., National Bioenergy Center, National Renewable Energy Laboratory
Beckham, G. T., National Renewable Energy Laboratory
Gill, R. T., University of Colorado Boulder

Synthetic Biology aims to implement new cellular features through the design of original genetic parts and systems. A wide variety of circuit architectures have been built successfully, proving the ability of synthetic genetic parts to predictably interact with each other within a single bacterial cell [4]. The next stage of synthetic biology will involve the *de novo* engineering of self-organizing multicellular systems. One of the main methods found in nature for creating complex spatially distributed systems is cellular differentiation through asymmetric cell division (ACD), whereby stem cells spawn off irreversibly differentiated daughters. The presence of different cell lineages within an eukaryotic tissue allows for the partitioning of metabolic and structural tasks. Furthermore, if one or both of the daughter cells becomes incapable of dividing, controlling asymmetric division probability can lead to tissue size and shape control [5]. In complex eukaryotic organisms ACD is typically brought about by the differential expression of key genes within each cell type. Here, we engineered a novel method for generating ACD in *Escherichia coli* and regulating its frequency within the population. This work provides a platform for the construction of complex synthetic tissues and organisms from bacterial cells that differentiate during development of the population.

Current methods for creating differentiated cell types in synthetic bacteria utilize gene regulatory networks that allow for bistable gene expression [2, 1]. While such systems generate bimodal populations, they are very sensitive to both intrinsic and extrinsic noise sources that can randomly force transitions between states [3]. We use a novel method for creating terminally differentiated synthetic bacterial cells. Specifically, we refactored DNA shuttling mechanisms from *Caulobacter crescentus*

into *E. coli* to target plasmid DNA into one and only one daughter cell upon division. Thus, the two daughter cells will not just have different gene expression states, but will be irreversibly genetically distinct. Our findings will lead to synthetic "tissues" that can be used in complex settings.

References

[1] R. G. Egbert and E. Klavins. Fine-tuning gene networks using simple sequence repeats.*Proceedings of the National Academy of Sciences*, 109(42):16817â□“16822, 2012.
[2] T. S. Gardner, C. R. Cantor, and J. J. Collins. Construction of a genetic toggle switch in *Escherichia coli*. *Nature*, 403(6767):339â□“342, 2000.
[3] C. Gupta, J. M. LÃ³pez, W. Ott, K. JosiÄ‡, and M. R. Bennett. Transcriptional delay stabilizes bistable gene networks. *Physical review letters*, 111(5):058104, 2013.
[4] A. S. Khalil and J. J. Collins. Synthetic biology: applications come of age. *Nature Reviews*

Genetics, 11(5):367â□“379, 2010.
[5] A. D. Lander, K. K. Gokoffski, F. Y. Wan, Q. Nie, and A. L. Calof. Cell lineages and the logic of proliferative control. *PLoS Biol*, 7(1):e1000015, 2009.

CRISPR NAND Gate: rescuing a cis-repressed gRNA using antisense sRNA

- Conference: Synthetic Biology Engineering Evolution Design SEED
- Year: 2017
- Proceeding: 2017 Synthetic Biology: Engineering, Evolution & Design (SEED)
- Group: Poster Session
- Session:

  Confirmed Posters

Authors:
Harvey, P. J., University of Edinburgh
Rosser, S. J., University of Edinburgh
Our ability to predict protein expression from DNA sequence alone remains poor, reflecting our limited understanding of cis-regulatory grammar and hampering the design of engineered genes for synthetic biology applications. We have generated a model that predicts the translational efficiency of the 5â□™ untranslated region (UTR) of mRNAs in the yeast *Saccharomyces cerevisiae.* We constructed a library of nearly half a million 50 nucleotide-long random 5â□™ UTRs and assayed them in a single massively parallel growth selection experiment. The resulting data have allowed us to quantify the impact on translation of Kozak sequence composition, upstream open reading frames (uORFs) and secondary structure. With this data, we have trained a convolutional neural network on the random library and validated it by predicting the translational efficiency of 5â□™ UTRs that natively occur in yeast. The model additionally was used to computationally evolve highly translating 5â□™ UTRs. We have confirmed experimentally that the great majority of the evolved sequences lead to higher translation rates than the starting sequences, demonstrating the predictive utility of this model.

Data Rich Strategies for Designing Cellular RNA Devices

- Conference: Synthetic Biology Engineering Evolution Design SEED
- Year: 2017
- Proceeding: 2017 Synthetic Biology: Engineering, Evolution & Design (SEED)
- Group: Poster Session
- Session:

  Confirmed Posters

Authors:
Xiang, J. S., Stanford University

Cell surface glycans make up an important mode of intercellular communication and the enzymes that assemble and degrade them are not only important targets for therapeutics, but also powerful tools for chemical glycobiology. Our recent work has explored 1) the use of engineered glycoside hydrolases for generating universal blood by antigen removal, 2) the identification of inhibitors to glycosyltransferases in cancer pathways, and 3) the development of engineered â□œglycosynthasesâ□□ to assemble and probe the function of glycosaminoglycan structures.

We have previously demonstrated that blood antigen-cleaving glycoside hydrolases can be engineered as tools towards the generation of universal blood through the cleavage of carbohydrate blood group antigens. Identification and directed evolution of these enzymes is enabled by a strategy that involves high-throughput screening with a sensitive fluorescence based assay. Using a similar approach, creatively repurposed to assay for glycosyltransferase rather than glycoside hydrolase activity, we have developed a high-throughput screen for inhibitors targeting fucosyltransferase enzymes that are highly expressed in cancers. Such inhibitors have potential as therapeutics for certain cancers wherein these enzymesâ□™ upregulated activity results in the assembly of carbohydrate structures (e.g. sialyl Lewis^X) involved in malignant processes including metastasis. In another focus, we are developing tools for chemoenzymatic glycan synthesis through mechanism-based engineering of â□œglycosynthaseâ□□ enzyme biocatalysts to assemble specific glycosaminoglycan structures. A facile enzymatic synthesis of these glycans will enable glycobiologists to probe their interactions in processes such as neural development and cancer.

Deep Device Mining for Carbohydrate-Active Enzymes in Pulp and Paper Mill Sludge Metagenome and Applications to Bioprocess Development

- Conference: Synthetic Biology Engineering Evolution Design SEED
- Year: 2017
- Proceeding: 2017 Synthetic Biology: Engineering, Evolution & Design (SEED)
- Group: Poster Session
- Session:

  Confirmed Posters

Authors:
Sharan, A. A., University of British Columbia
Yadav, V. G., University of British Columbia
Hallam, S. J., University of British Columbia

Biocatalyst discovery is integral to bioecomony development, enabling design of robust and scalable bioprocesses that are able to compete with the resource intensive petrochemical industry. Uncultivated microbial communities within natural and engineered ecosystems provide a near-infinite reservoir of genomic diversity and metabolic potential that can be harnessed for biocatalyst discovery[1]. To bridge the cultivation gap, functional metagenomic screens have been developed to recover active genes directly from environmental samples[2]. Here, we describe a pipeline for recovery of biomass-deconstructing biocatalysts sourced from pulp and paper mill sludge metagenomes to develop consolidated bioprocesses that convert waste streams into value-added products. High molecular weight DNA was extracted from sludge and used to construct a fosmid library containing 15,000 clones using the copy control system in EPI300â„¢-T1 RÂ *E. coliÂ* based on established robotic methods[3]. Extracted DNA was also used in whole genome shotgun sequencing to compare the metabolic potential of the sludge community with fosmid screening outcomes using MetaPathways[4]Â with specific emphasis on carbohydrate-active enzymes (CAZymes). In total, 32,232 ORFâ□™s hits in CAZy database were obtained including glycoside hydrolases, glycosyl transferases, and carbohydrate binding module families. The fosmid library was screened for glycosidase hydrolase activities using a pool of sensitive fluorogenic glycosides of 6-chloro-4-methylumbelliferoneÂ enabling rapid recovery of Î²-cellobiosidase, mannosidase and xylosidase activitiesÂ 5. A total of 744 clones capable of converting pooled substrates were recovered indicating an extremely high hit rate (1 hit per 43 clones). Downstream characterization including complete fosmid sequencing and subcloning results to identify specific active genes or gene cassettes is currently in progress. These results reinforce the notion that screening libraries sourced from environments naturally enriched for lignocellulosic biomass deconstruction increases screening efficieny and chances of recovering biocatalysts optimized to perform under specific environmental conditions. The combination of robotic library production and pooled screening using enriched DNA sources for lignocellulosic biomass deconstruction provides an example of high throughput experimental design extensible to a wide range of screening targets.

1. Mewis, K., Armstrong, Z., Song, Y. C., Baldwin, S. A., Withers, S. G., & Hallam, S. J. (2013). Biomining active cellulases from a mining bioremediation system. *Journal of biotechnology*, *167*(4), 462-471.
2. Nyyssönen, M., Tran, H. M., Karaoz, U., Weihe, C., Hadi, M. Z., Martiny, J. B., ... & Brodie, E. L. (2013). Coupled high-throughput functional screening and next generation sequencing for identification of plant polymer decomposing enzymes in metagenomic libraries. *Frontiers in microbiology*, *4*, 282.
3. Mewis, K., Taupp, M., & Hallam, S. J. (2011). A high throughput screen for biomining cellulase activity from metagenomic libraries. *JoVE (Journal of Visualized Experiments)*, (48), e2461-e2461.
4. Hanson, N. W., Konwar, K. M., Wu, S. J., & Hallam, S. J. (2014, May). MetaPathways v2. 0: A master-worker model for environmental Pathway/Genome Database construction on grids and clouds. In *Computational Intelligence in Bioinformatics and Computational Biology, 2014 IEEE Conference on*(pp. 1-7). IEEE.
5. Chen, Hong-Ming, et al. "Synthesis and evaluation of a series of 6-chloro-4-methylumbelliferyl glycosides as fluorogenic reagents for screening metagenomic libraries for glycosidase activity." Carbohydrate research 421 (2016): 33-39.

Poster #132: Deep Device Mining for Carbohydrate-Active Enzymes in Pulp and Paper Mill Sludge Metagenome and Applications to Bioprocess Development

- Anupama A Sharan
University of British Columbia, Vancouver

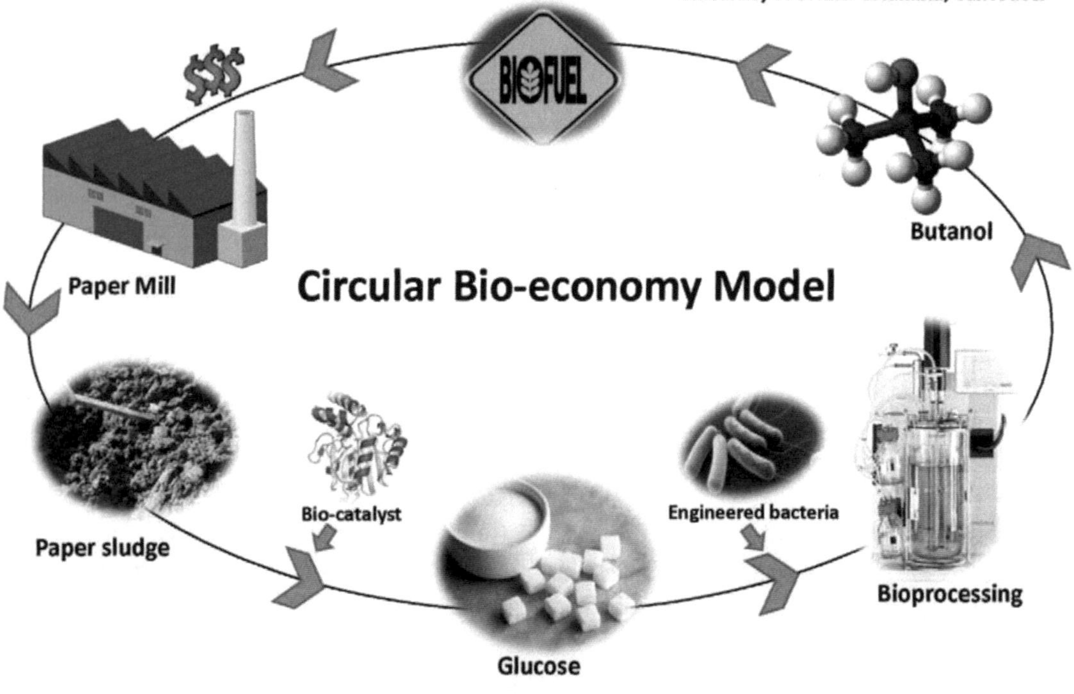

Deep Device Mining for Lignin-Transforming Enzymes Using Whole-Cell Biosensors

- Conference: Synthetic Biology Engineering Evolution Design SEED
- Year: 2017
- Proceeding: 2017 Synthetic Biology: Engineering, Evolution & Design (SEED)
- Group: Poster Session
- Session:

  Confirmed Posters

Authors:
Ho, J. C. H., University of British Columbia
Hallam, S. J., University of British Columbia
Pawar, S. V., University of British Columbia
Yadav, V. G., University of British Columbia

A genetic toggle switch is a circuit that possesses two stable opposite states of gene expression. Many synthetic circuits that are applied to biotechnology are based on toggle switches, such as biosensors. However, extrinsic and intrinsic noise can cause random switching and generate a variety of gene expression states. To make circuits more robust, we propose the creation of a consensus toggle switch that will have the ability to amplify induction signals among neighboring cells and achieve a consensus in the population. We will create the consensus by employing quorum sensing in our circuits. Quorum sensing (QS) is the mechanism through which bacteria communicate with each other, by producing signaling molecules that can affect gene expression in a density-dependent way. We hypothesize that synthetic toggles that coordinate their behavior with QS can better perform tasks that are difficult for uncoordinated systems. Here, we constructed 4 versions of consensus toggles to understand how the presence/absence of positive feedback loops in quorum-sensing genes and direct/indirect repression affect toggle behavior. In our designs, each possible state of the toggle is reported by one color: yellow (YFP) or blue (CFP). We find that the dynamics of consensus toggle switches depend on their regulatory topologies. Topologies that contain indirect repression show a peak of YFP expression followed by a shutdown, while topologies with direct repression do not shut down. Furthermore, topologies with indirect repression do not show distinguishable CFP expression between induced and uninduced samples, while topologies with direct repression do. Topologies that lack extra feedback loops show less variability, while topologies that contain these extra feedback loops are more variable. These results enable the selection of a topology that works better for the creation of a consensus toggle. In the microscope, the toggle with direct repression and without extra feedback loops showed faster switch compared to a conventional toggle. These results are encouraging, because a consensus toggle has the potential of generating more sensitive and robust detection of signals than regular toggles. And to further apply this circuit, such as for the detection of diseases/infections in the mammalian gastrointestinal tract or for the surveillance and bioremediation of environments, it is desired that the circuit shows sensitivity and robustness.

Deep Mutational Scanning of NDM-1: Reverse Engineering Antibiotic Resistance to Decode the Protein Sequence and Function Relationship

- Conference: Synthetic Biology Engineering Evolution Design SEED
- Year: 2017
- Proceeding: 2017 Synthetic Biology: Engineering, Evolution & Design (SEED)
- Group: Poster Session
- Session:

  Confirmed Posters

Authors:
Life, B. E., University of British Columbia
Chen, J., UBC
Socha, R., University of British Columbia
Tokuriki, N., University of British Columbia
Fowler, D., University of Washington

CRISPR/Cas9 has wide-reaching scientific and therapeutic applications, but efficient, exact targeting of cutting sites can be challenging. Several cas9 variants have been engineered to target alternate PAMs; however, generation of new variants is often laborious. Additionally, the cutting ability of cas9 nucleases is often highly guide-dependent and shows a large amount of variation between sites. In order to engineer new nuclease variants targeted to cleave a given site, we developed a directed evolution platform to select for DNA cleavage in bacteria. Our system allows rapid generation of mutant libraries, which are challenged with phage containing target sites in a competitive pool. Using this method, we developed *S. aureus* cas9 variants which have altered PAM preferences to target cleavage to novel targets of therapeutic relevance.

Developing HIV-1 Vaccine through a Synthetic Biology Approach

- Conference: Synthetic Biology Engineering Evolution Design SEED
- Year: 2017
- Proceeding: 2017 Synthetic Biology: Engineering, Evolution & Design (SEED)
- Group: Poster Session
- Session:

  Confirmed Posters

Authors:
Niu, W., University of Nebraska - Lincoln
Wang, N.
Yuan, Z.
Li, Q.
Guo, J.

An effective vaccine for human immunodeficiency virus type-1 (HIV-1) is highly desirable to stem worldwide AIDS pandemic. Although live-attenuated virus vaccine showed excellent efficacy in eliciting hostÃ¢â‚¬â„¢s immune response, application of such strategy in HIV-1 prevention is hindered by safety concerns due to uncontrolled replication of virus. We recently introduced a new synthetic biology approach to control HIV-1 replication, which entails the manipulation of essential HIV-1 protein biosynthesis through nonsense codon reassignment. This is achieved by the insertion of one or more amber nonsense codons into essential genes of HIV-1. In the presence of an engineered amber suppression machinery, the amber nonsense codon is reassigned into a sense codon to encode an unnatural amino acid. We have successfully demonstrated that we could turn HIV-1 replication on and off in vitro using an unnatural amino acid as the cue. We also engineered HIV-1 genome to contain a genomic copy of the amber suppression machinery, which enables multi-cycle replication control and immunological evaluation in vivo. We envisage that allowing live-attenuated HIV-1 vaccine to replicate only in the presence of unnatural amino acid would improve its safety while still maintains its good efficacy against HIV-1 acquisition.Â This new synthetic biology approach could be potentially applied to the generation of vaccines against other pathogenic viruses or bacteria.

Developing HIV-1 Vaccine through a Synthetic Biology Approach

Wei Niu,*[1] Nanxi Wang,[2] Zhe Yuan,[3] Qingsheng Li,*[3] Jiantao Guo*[2]

1. Department of Chemical & Biomolecular Engineering, 2. Department of Chemistry, 3. Nebraska Center for Virology
University of Nebraska-Lincoln, Lincoln, Nebraska, 68588, USA.

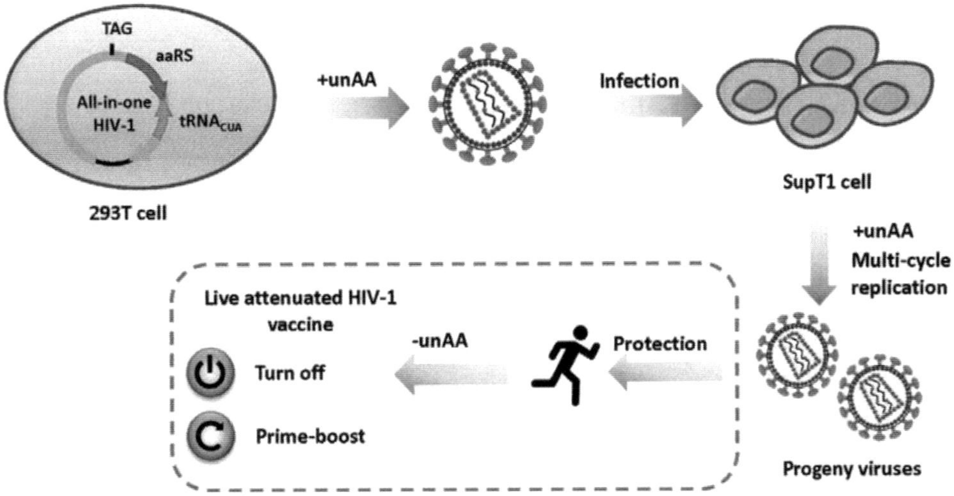

University of Nebraska Lincoln

Poster #46

Developing Synthetic Biology Capabilities in the Fast-Growing Bacterium Vibrio Natriegens

- Conference: Synthetic Biology Engineering Evolution Design SEED
- Year: 2017
- Proceeding: 2017 Synthetic Biology: Engineering, Evolution & Design (SEED)
- Group: Poster Session
- Session:

  Confirmed Posters

Authors:
Tschirhart, T., Naval Research Laboratory
Wang, Z., Naval Research Laboratory
Erickson, J., Naval Research Laboratory
Vora, G., Naval Research Laboratory

CRISPR/Cas9 has wide-reaching scientific and therapeutic applications, but efficient, exact targeting of cutting sites can be challenging. Several cas9 variants have been engineered to target alternate PAMs; however, generation of new variants is often laborious. Additionally, the cutting ability of cas9 nucleases is often highly guide-dependent and shows a large amount of variation between sites. In order to engineer new nuclease variants targeted to cleave a given site, we developed a directed evolution platform to select for DNA cleavage in bacteria. Our system allows rapid generation of mutant libraries, which are challenged with phage containing target sites in a competitive pool. Using this method, we developed *S. aureus* cas9 variants which have altered PAM preferences to target cleavage to novel targets of therapeutic relevance.

Development and Substrate Specificity Screening of an In Vivo Biosensor for the Detection of Biomass Derived Aromatic Chemical Building Blocks

- Conference: Synthetic Biology Engineering Evolution Design SEED
- Year: 2017
- Proceeding: 2017 Synthetic Biology: Engineering, Evolution & Design (SEED)
- Group: Poster Session
- Session:

  Confirmed Posters

Authors:
Machado, L. F. M., University of Manchester
Dixon, N.

State-of-the-art strain engineering techniques for the protein producing yeast host *Pichia pastoris* include overexpression of homologous and heterologous genes, and deletion of host genes. For this purpose overexpression vectors and gene deletion methods such as the split marker technique have been established. For metabolic and cell engineering purposes, the simultaneous overexpression of more than one gene is often needed. Previous approaches employing subsequent steps of overexpression and marker recycling were time- and labor-consuming. Therefore, efficient systems allowing multiple gene overexpression are required, that can be stably integrated into the *P. pastoris* genome. To this end, we developed a synthetic biology toolbox based on Golden Gate cloning to enable efficient construction of complex and versatile over-expression vectors. Up to five different expression cassettes, employing a library of promoters and terminators can be combined into one vector, and successfully integrated into thc genomic DNA of *P. pastoris* at targeted loci in one step. Recent trends in synthetic biology, however, go into the direction of building up large and complex reaction networks. To allow for clean and unscarred genetic engineering, a CRISPR/Cas9 based method for gene insertions, deletions and replacements was developed, which paves the way for precise genomic rearrangements in *P. pastoris*. By using this technique precise genomic integrations were performed efficiently without integrative selection markers. The repertoire of genetic techniques developed so far, will provide a wide variety of possibilities to engineer *P. pastoris*. Applications for these synthetic biology tools in cell engineering of recombinant *P. pastoris* will be presented.

Development of a Conjugation-Based Genome Editing Tool in Shewanella Oneidensis MR-1 Using CRISPR/Cas9 and Recombineering

- Conference: Synthetic Biology Engineering Evolution Design SEED
- Year: 2017
- Proceeding: 2017 Synthetic Biology: Engineering, Evolution & Design (SEED)
- Group: Poster Session
- Session:

  Confirmed Posters

Authors:
Domènech Corts, A., University of Colorado
Gill, R. T., University of Colorado Boulder
Gralnick, J. A., University of Minnesota

Synthetic Biology aims to implement new cellular features through the design of original genetic parts and systems. A wide variety of circuit architectures have been built successfully, proving the ability of synthetic genetic parts to predictably interact with each other within a single bacterial cell [4]. The next stage of synthetic biology will involve the *de novo* engineering of self-organizing multicellular systems. One of the main methods found in nature for creating complex spatially distributed systems is cellular differentiation through asymmetric cell division (ACD), whereby stem cells spawn off irreversibly differentiated daughters. The presence of different cell lineages within an eukaryotic tissue allows for the partitioning of metabolic and structural tasks. Furthermore, if one or both of the daughter cells becomes incapable of dividing, controlling asymmetric division probability can lead to tissue size and shape control [5]. In complex eukaryotic organisms ACD is typically brought about by the differential expression of key genes within each cell type. Here, we engineered a novel method for generating ACD in *Escherichia coli* and regulating its frequency within the population. This work provides a platform for the construction of complex synthetic tissues and organisms from bacterial cells that differentiate during development of the population.

Current methods for creating differentiated cell types in synthetic bacteria utilize gene regulatory networks that allow for bistable gene expression [2, 1]. While such systems generate bimodal populations, they are very sensitive to both intrinsic and extrinsic noise sources that can randomly force transitions between states [3]. We use a novel method for creating terminally differentiated synthetic bacterial cells. Specifically, we refactored DNA shuttling mechanisms from *Caulobacter crescentus*

into *E. coli* to target plasmid DNA into one and only one daughter cell upon division. Thus, the two daughter cells will not just have different gene expression states, but will be irreversibly genetically distinct. Our findings will lead to synthetic "tissues" that can be used in complex settings.

References

[1] R. G. Egbert and E. Klavins. Fine-tuning gene networks using simple sequence repeats.*Proceedings of the National Academy of Sciences*, 109(42):16817â□“16822, 2012.

[2] T. S. Gardner, C. R. Cantor, and J. J. Collins. Construction of a genetic toggle switch in *Escherichia coli*. *Nature*, 403(6767):339â□“342, 2000.

[3] C. Gupta, J. M. LÃ³pez, W. Ott, K. JosiÄ‡, and M. R. Bennett. Transcriptional delay stabilizes bistable gene networks. *Physical review letters*, 111(5):058104, 2013.

[4] A. S. Khalil and J. J. Collins. Synthetic biology: applications come of age. *Nature Reviews*

Genetics, 11(5):367â□“379, 2010.

[5] A. D. Lander, K. K. Gokoffski, F. Y. Wan, Q. Nie, and A. L. Calof. Cell lineages and the logic of proliferative control. *PLoS Biol*, 7(1):e1000015, 2009.

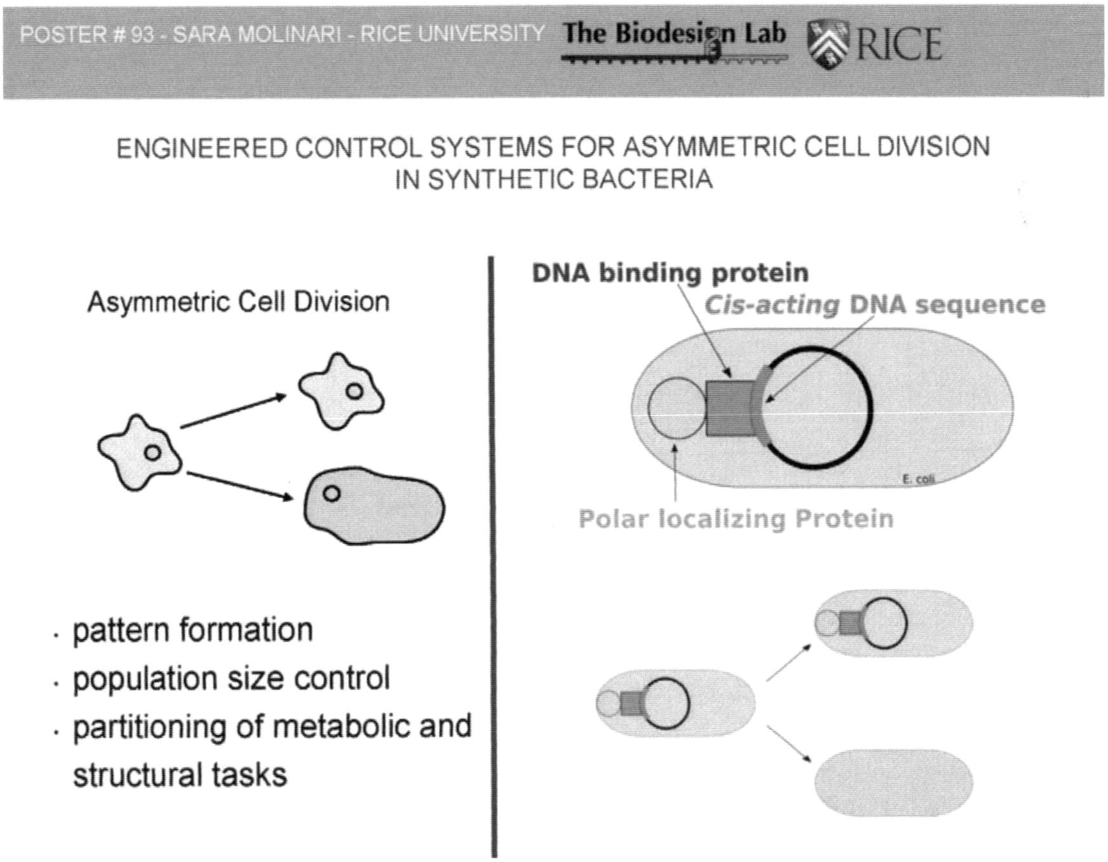

Development of a Fast-Responding, Minimal Equipment Biosensor for Zinc Deficiency

- Conference: Synthetic Biology Engineering Evolution Design SEED
- Year: 2017
- Proceeding: 2017 Synthetic Biology: Engineering, Evolution & Design (SEED)
- Group: Poster Session
- Session:

  Confirmed Posters

Authors:
McNerney, M., Georgia Institute of Technology
Styczynski, M. P., Georgia Institute of Technology

Zinc deficiency contributes to hundreds of thousands of childhood deaths annually, but current detection methods are cost-prohibitive for use in the regions most affected. Our group recently developed a low-cost bacterial biosensor that can produce different visible outputs based on the concentration of zinc in which the cells grow. However, these initial sensor cells always produce some pigment, which necessitates that they be grown from a very small inoculum so that initial coloration is minimal â" thus requiring overnight assay times. To make an easy-to-use, field-deployable assay, we are engineering cells that completely repress pigmentation during growth and upon induction quickly produce different pigments based on zinc concentration. This would allow sample addition to concentrated pre-cultured cells, decreasing assay time from about 24 hours to less than 3 hours.

We first focused on reducing unwanted pigment expression and enabling fast production of the red pigment lycopene upon induction. A sensor that produces a pigment instead of fluorescent proteins is advantageous in resource-limited settings, since sensor readout can be interpreted without advanced equipment. However, complete repression of pigment production requires more precise control than what is necessary for most synthetic biology applications, since small amounts of enzyme can produce visible amounts of pigment, (especially over long times as may be necessary to pre-culture sensor cells). To combat unwanted pigment expression, three commonly used inducible promoter systemsâ"the *lac*, *ara*, and *T7* systemsâ"were engineered to be more repressible. Despite orders of magnitude decreases in uninduced protein expression, even the most repressible systems still could not control lycopene production. Translational modifications proved much more effective at controlling lycopene production. When the original ribosomal binding sites for the *crtE*, *crtB*, and *crtI* genes were replaced with weak RBSs, all systems could fully repress lycopene production and produce visible lycopene within three hours of induction. Supplementation of metabolic precursor pathways further reduced the time required for visible pigmentation to 1.5 hours.

To engineer a zinc-sensitive response, we next incorporated zinc-responsive elements into the engineered promoter systems. Operator sites for the zinc-responsive transcription factor Zur were introduced into *lac*, *ara*, and *T7* promoter systems to create systems that respond both to an exogenous inducer (either IPTG or arabinose) and to zinc. Promoters from all groups demonstrated a response to both exogenous inducer and zinc, and the best responding promoter was a T7-based promoter with binding sites for Zur and LacI. These promoter systems were used to control expression of violacein, and transcriptional switches and inducible protein degradation were explored as methods to increase the dynamic range of the sensor. Additional zinc-responsive promoters and pigment pathways can be added to create a multi-color litmus test for zinc concentration.

To engineer a system that produces specific pigments based on external state, translational control eliminated unwanted pigment expression, and transcriptional control enabled dual regulation. As the applications of synthetic biology expand to systems that require more precise control of metabolic state, multiple layers of regulation will likely be necessary to engineer the desired output.

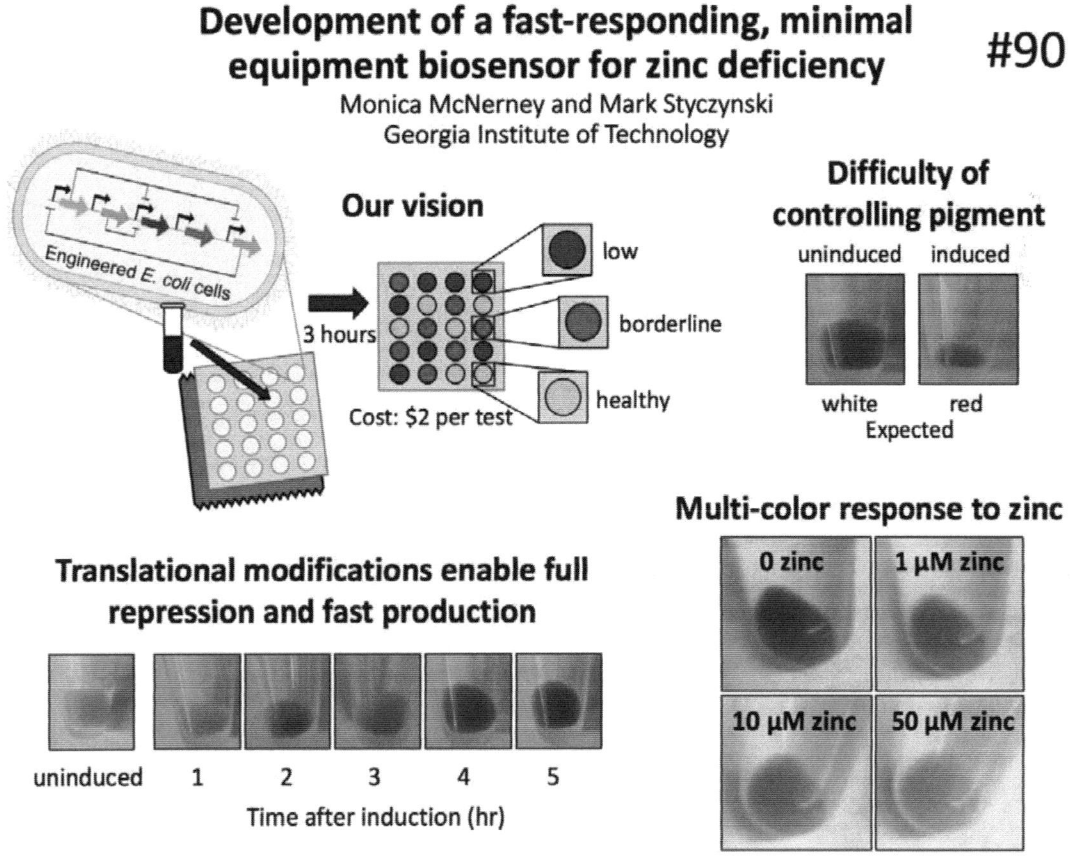

Development of a Microbial Biosensor for Detection of Inflammation-Associated Microenvironments

- Conference: Synthetic Biology Engineering Evolution Design SEED
- Year: 2017
- Proceeding: 2017 Synthetic Biology: Engineering, Evolution & Design (SEED)
- Group: Poster Session
- Session:

  Confirmed Posters

Authors:
Galley, J.
Tabor, J. J., Rice University
Chazin, W., Vanderbilt University
Britton, R.
Daeffler, K., Rice University
DNA assembly forms the cornerstone of modern synthetic biology. Despite the numerous available methods, scarless multi-fragment assembly of large plasmids remains challenging. Furthermore, the upcoming wave in molecular biological automation demands a rethinking of how we perform DNA assembly. To streamline automation workflow and minimize operator intervention, a non-enzymatic assembly method is highly desirable. Here, we report the optimization and operationalization of a process called Twin-Primer Assembly (TPA), which is a method to assemble PCR-amplified fragments into a plasmid without the use of enzymes.

Each fragment used in TPA requires two PCR products, amplified separately using two sets of primers. One product has an overlap to the fragment that comes before and the other has an overlap to the fragment that comes after. The length of the overlap region is determined by the melting temperature (T_M) of the overlap region, and all overlaps used in an assembly are designed to have the same T_M. The two PCR products are mixed and fully denatured at high temperature before they are allowed to re-anneal slowly, generating intermediates that have overhangs on both sides. These intermediates are then allowed to hybridize at an elevated temperature to form a circular nicked plasmid. Upon *E. coli* transformation, the cells phosphorylate and ligate the nicks to create the final product plasmid.

The TPA non-enzymatic DNA assembly method is scarless and generally sequence independent. It is capable of assembling a 7 kb plasmid from 10 fragments at close to 80% fidelity, and a 31kb plasmid from 5 fragments at around 50% fidelity. Assembling more than 10 fragments is likely possible for constructing smaller plasmids, while assembling a plasmid larger than 31 kb is likely possible using fewer fragments. Even without the use of enzymes, the performance of TPA is on par with some of the best *in vitro* assembly methods currently available. TPA should be an invaluable addition to a synthetic biologistâ□™s toolbox.

Development of a Xylose Utilizing Saccharomyces Cerevisiae As a Platform for Lignocellulosic Biofuels/Biochemical Production

- Conference: Synthetic Biology Engineering Evolution Design SEED
- Year: 2017
- Proceeding: 2017 Synthetic Biology: Engineering, Evolution & Design (SEED)
- Group: Poster Session
- Session:

  Confirmed Posters

Authors:
Lee, S. M., Korea Institute of Science and Technology
Tran, P. H. N., Korea University of Science and Technology
Amyris is an end-to-end biological manufacturer that uses synthetic biology to engineer microbes as catalysts for conversion of sugar into chemical products. Many of these chemical targets were previously derived from limited natural resources and economically inaccessible via traditional chemistries. In the last decade, Amyris has developed a sophisticated technology platform for engineering microbes to produce high levels of isoprenoids including artemisinin and farnesene. In recent years, we have begun to harness the power of this proven infrastructure to expand the portfolio of bio-manufactured chemical products.

In this talk, I will describe the current state-of-the-art tools at Amyris, including our automated strain engineering, high-throughput screening, and fermentation. Then I will discuss how these tools are evolving under the DARPA-funded â□œMgs to Kgsâ□□ project to enable the concurrent development of 100â□™s of products simultaneously. In the first year, the project has produced more than 150 target molecules from all parts of metabolic space. As the project continues, â□œMgs to Kgsâ□□ will enable Amyris to incorporate next generation high-throughput â□˜omics, automation, data science, and machine learning tools to further increase the rate at which biology can be engineered.

Directed Combinatorial Mutagenesis of Escherichia coli for Complex Phenotype Engineering

- Conference: Synthetic Biology Engineering Evolution Design SEED
- Year: 2017
- Proceeding: 2017 Synthetic Biology: Engineering, Evolution & Design (SEED)
- Group: Poster Session
- Session:

  Confirmed Posters

Authors:
Liu, R., University of Colorado Boulder
Gill, R. T., University of Colorado Boulder
Choudhury, A., University of Colorado, Boulder
Beckham, G. T., National Renewable Energy Laboratory
Liang, L., University of Colorado Boulder
Nogué, V. S. I., National Bioenergy Center, National Renewable Energy Laboratory
Garst, A., Muse Biotechnology Inc.
Mosquito-borne diseases such as Zika, dengue fever and Chikungunya are major and increasing international public health concerns. The primary vector of these viral diseases is *Aedes aegypti* and current control measures against this mosquito are proving inadequate or problematic. An innovative and effective control technique for *Aedes aegypti* involves the release of self-limiting mosquitoes. These mosquitoes are genetically engineered to be homozygous for a self-limiting gene under control of an on/off tetracycline switch. When released into the wild, any offspring of the transgenic males will inherit one copy of the transgene. In the absence of tetracycline antidote, the transgene will prevent the mosquito from developing to adulthood, thus suppressing the population with repeated releases. Oxitecâ□™s self-limiting transgenic insect technology has also been applied successfully to develop population control measures for major crop pests such as the Mediterranean fruit fly (Medfly) and the diamondback moth. In a variation on the self-limiting genetics used in *Aedes* mosquitoes, genetic sexing strains that express the self-limiting gene in a sex-specific manner are used to produce male-only cohorts of these crop pests for release in control programmes. These and other developments in Oxitecâ□™s genetic engineering programme to enable biological control of major insect pests will be discussed. The Oxitec transgenic mosquito method has dramatically reduced *Aedes aegypti* populations in trials around the world and is now being used under operational conditions in the Cayman Islands and Brazil to control *Aedes aegypti*. Results of these programmes will be presented.

Directed Evolution of an Array of Orthogonal Biosensors

- Conference: Synthetic Biology Engineering Evolution Design SEED
- Year: 2017
- Proceeding: 2017 Synthetic Biology: Engineering, Evolution & Design (SEED)
- Group: Poster Session
- Session:

  Confirmed Posters

Authors:
Meyer, A. J., Massachusetts Institute of Technology
Voigt, C. A., Massachusetts Institute of Technology

Small molecule-inducible transcription factors, or sensors, enable temporal control of toxic genes, provide the inputs forÂ *in vivo*Â computation, and allow a host to sense and respond to its environment. In this ongoing work we endeavor to improve the characteristics of several sensors using a mixture of rational engineering and directed evolution. We have improved the dynamic range, sensitivity, and/or specificity of 18 sensors. As an example of the utility of these sensors, we have used five optimized sensors to independently control the expression of five components of a metabolic pathway. By varying the amount of each inducer, we can easily and rapidly optimize the expression profile of the pathway without the need to construct large libraries. To facilitate the adoption of multiplex, sensor-based genetic control, we have created a stable cell-line capable of responding to 11 inputs, thus enabling the modular control of multiple genetic outputs with minimal optimization.

Directed evolution of an array of orthogonal biosensors

Adam Meyer (Voigt Lab) - Poster 92

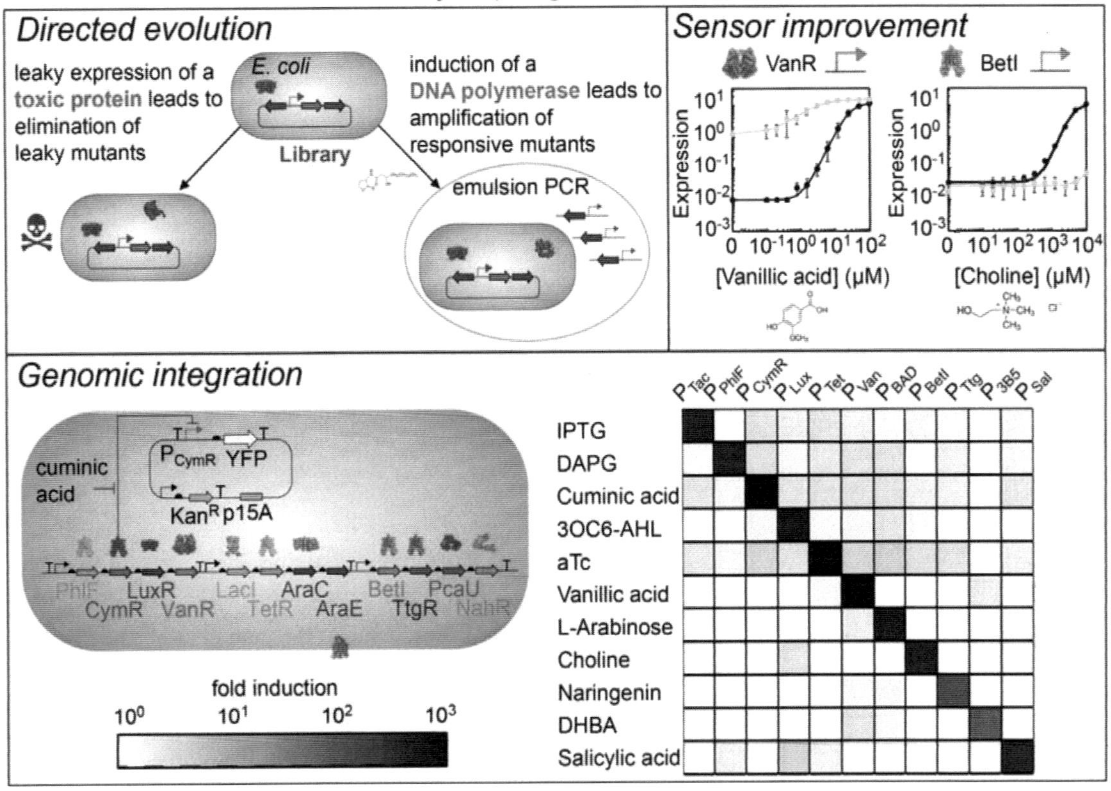

Dissecting the Factors That Determine the Fidelity of the Genetic Code through Sense Codon Reassignment

- Conference: Synthetic Biology Engineering Evolution Design SEED
- Year: 2017
- Proceeding: 2017 Synthetic Biology: Engineering, Evolution & Design (SEED)
- Group: Poster Session
- Session:

  Confirmed Posters

Authors:
Schmitt, M. A., Colorado State University
Fisk, J. D., Colorado State University
Schwark, D., Colorado State University

Plants are a rich source of unique scaffolds, including 25% of natural-product-derived drugs. However, the discovery, synthesis, and overall material supply chains for sourcing plant natural products and their derivatives remain ad hoc, biased, and tedious. While microbial biosynthesis presents compelling alternatives to traditional approaches based on extraction from natural plant hosts, many challenges exist in the reconstruction of plant specialized metabolic pathways in microbial hosts. My laboratory has developed approaches to address the challenges that arise in the reconstruction of complex biosynthesis schemes, including spatial engineering strategies to direct the activities and specificities of pathway enzymes and recoding strategies to address folding, processing, and stability issues that may arise with the expression of plant enzymes in heterologous microbial hosts. We have utilized these strategies to develop yeast-based production platforms for an important class of plant alkaloids, the benzylisoquinoline alkaloids, including the medicinal opioids. These synthetic biology platforms will lead to transformative advances in natural product discovery, drug development, and production.

DNA Construction Using DIVA (Design, Implementation, Validation Automation): Wet-Lab Workflows and Software Platform

- Conference: Synthetic Biology Engineering Evolution Design SEED
- Year: 2017
- Proceeding: 2017 Synthetic Biology: Engineering, Evolution & Design (SEED)
- Group: Poster Session
- Session:

  Confirmed Posters

Authors:
Kaplan, N., Joint BioEnergy Institute
Goyal, G., Joint BioEnergy Institute
Chiniquy, J., DOE Agile BioFoundry
Plahar, H., Joint BioEnergy Institute
Chen, J., Joint BioEnergy Institute, Lawrence Berkeley National Laboratory
Tapaswi, M., Joint BioEnergy Institute, Lawrence Berkeley National Laboratory
Stawski, N., Lawrence Berkeley National Laboratory
Simirenko, L., DOE Joint Genome Institute, Lawrence Berkeley National Laboratory
Deutsch, S., Lawrence Berkeley National Laboratory
Guenther, J. M., 5Sandia National Laboratory
Hillson, N. J., Joint BioEnergy Institute

Synthetic Biology aims to implement new cellular features through the design of original genetic parts and systems. A wide variety of circuit architectures have been built successfully, proving the ability of synthetic genetic parts to predictably interact with each other within a single bacterial cell [4]. The next stage of synthetic biology will involve the *de novo* engineering of self-organizing multicellular systems. One of the main methods found in nature for creating complex spatially distributed systems is cellular differentiation through asymmetric cell division (ACD), whereby stem cells spawn off irreversibly differentiated daughters. The presence of different cell lineages within an eukaryotic tissue allows for the partitioning of metabolic and structural tasks. Furthermore, if one or both of the daughter cells becomes incapable of dividing, controlling asymmetric division probability can lead to tissue size and shape control [5]. In complex eukaryotic organisms ACD is typically brought about by the differential expression of key genes within each cell type. Here, we engineered a novel method for generating ACD in *Escherichia coli* and regulating its frequency within the population.

This work provides a platform for the construction of complex synthetic tissues and organisms from bacterial cells that differentiate during development of the population.

Current methods for creating differentiated cell types in synthetic bacteria utilize gene regulatory networks that allow for bistable gene expression [2, 1]. While such systems generate bimodal populations, they are very sensitive to both intrinsic and extrinsic noise sources that can randomly force transitions between states [3]. We use a novel method for creating terminally differentiated synthetic bacterial cells. Specifically, we refactored DNA shuttling mechanisms from *Caulobacter crescentus*

into *E. coli* to target plasmid DNA into one and only one daughter cell upon division. Thus, the two daughter cells will not just have different gene expression states, but will be irreversibly genetically distinct. Our findings will lead to synthetic "tissues" that can be used in complex settings.

References

[1] R. G. Egbert and E. Klavins. Fine-tuning gene networks using simple sequence repeats.*Proceedings of the National Academy of Sciences*, 109(42):16817â□“16822, 2012.
[2] T. S. Gardner, C. R. Cantor, and J. J. Collins. Construction of a genetic toggle switch in *Escherichia coli*. *Nature*, 403(6767):339â□“342, 2000.
[3] C. Gupta, J. M. LÃ³pez, W. Ott, K. JosiÄ‡, and M. R. Bennett. Transcriptional delay stabilizes bistable gene networks. *Physical review letters*, 111(5):058104, 2013.
[4] A. S. Khalil and J. J. Collins. Synthetic biology: applications come of age. *Nature Reviews*

Genetics, 11(5):367â□“379, 2010.
[5] A. D. Lander, K. K. Gokoffski, F. Y. Wan, Q. Nie, and A. L. Calof. Cell lineages and the logic of proliferative control. *PLoS Biol*, 7(1):e1000015, 2009.

Easyclone-Markerfree: A Vector Toolkit for Marker-Less Integration of Genes into Saccharomyces Cerevisiae Via CRISPR-Cas9

- Conference: Synthetic Biology Engineering Evolution Design SEED
- Year: 2017
- Proceeding: 2017 Synthetic Biology: Engineering, Evolution & Design (SEED)
- Group: Poster Session
- Session:

  Confirmed Posters

Authors:
jessop Fabre, M., DTU Biosustain
Jakociunas, T., The Novo Nordisk Foundation Center for Biosustainability
Stovicek, V., Technical University of Denmark
Dai, Z., Chalmers University of Technology
Jensen, M. K., Technical University of Denmark
Keasling, J., University of California, Berkeley
Borodina, I., Technical University of Denmark

Bacteria lack compartmentalization and therefore rely on subcellular organization to temporally and spatially localize major cellular processes. As one of the major structures in the cell, the nucleoid possesses positional cues that help direct important functions such as DNA replication and cell division. While some of the factors that contribute to maintaining subcellular organization are characterized, many of the underlying determinants are still unknown. In *Bacillus subtilis*, there remains a class of putative DNA-binding proteins currently uncharacterized, and I hypothesize that some of these proteins use DNA to localize functions important in maintaining subcellular organization. Since many uncharacterized genes lack easily discernable knockout phenotypes, I utilized a gene misexpression approach to identify DNA-binding proteins capable of perturbing some aspect of cellular organization. The misexpression library employed comprises over 800 strains, each containing an uncharacterized gene under the control of an inducible promoter on the *B. subtilis* chromosome. Genes for 98 putative DNA-binding proteins were misexpressed and analyzed for their ability to grow in the presence of inducer. Of these strains, nine exhibited reduced growth and 17 strains did not grow. When analyzed by fluorescence microscopy, 19 of these strains could be associated with a distinct phenotype, including defects in the structure and placement of the chromosome, abnormal cell division, and loss of cell shape. To further characterize strains with the most severe phenotypes we performed RNA-seq experiments to determine if gene misexpression alters the transcription profile. We identified four genes that do not affect transcription, suggesting that the misexpression phenotype is not due to transcriptional regulation. We hypothesize that this class of DNA-binding proteins act as positional cues that help maintain subcellular organization.

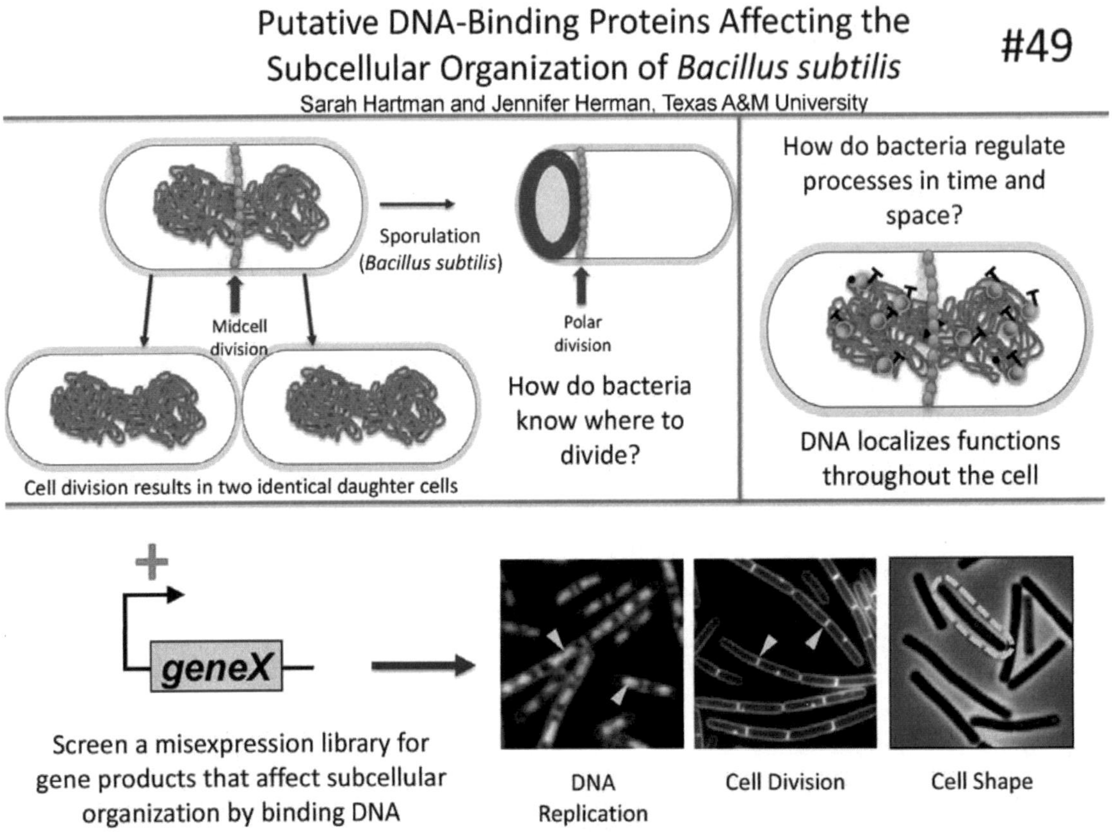

Effective Miniaturization of Illumina Nextera XT Library Preparation for Small Whole Genome Sequencing By Utilizing the Labcyte Echo 525® Liquid Handler

- Conference: Synthetic Biology Engineering Evolution Design SEED
- Year: 2017
- Proceeding: 2017 Synthetic Biology: Engineering, Evolution & Design (SEED)
- Group: Poster Session
- Session:

  Confirmed Posters

Authors:
Lai, J., Amyris
Lesnick, J., Labcyte
Ruppert, N.

Plants are a rich source of unique scaffolds, including 25% of natural-product-derived drugs. However, the discovery, synthesis, and overall material supply chains for sourcing plant natural products and their derivatives remain ad hoc, biased, and tedious. While microbial biosynthesis presents compelling alternatives to traditional approaches based on extraction from natural plant hosts, many challenges exist in the reconstruction of plant specialized metabolic pathways in microbial hosts. My laboratory has developed approaches to address the challenges that arise in the reconstruction of complex biosynthesis schemes, including spatial engineering strategies to direct the activities and specificities of pathway enzymes and recoding strategies to address folding, processing, and stability issues that may arise with the expression of plant enzymes in heterologous microbial hosts. We have utilized these strategies to develop yeast-based production platforms for an important class of plant alkaloids, the benzylisoquinoline alkaloids, including the medicinal opioids. These synthetic biology platforms will lead to transformative advances in natural product discovery, drug development, and production.

Effects of Overexpressing Calvin-Bassham-Benson Cycle Genes in Synechocystis PCC 6803

- Conference: Synthetic Biology Engineering Evolution Design SEED
- Year: 2017
- Proceeding: 2017 Synthetic Biology: Engineering, Evolution & Design (SEED)
- Group: Poster Session
- Session:

  Confirmed Posters

Authors:
Liang, F., Uppsala University
Lindblad, P., Uppsala University
3-hydroxypropanoic acid (3-HP) is an important biomass-derivable platform chemical that can be converted into a number of industrially relevant compounds. There have been several attempts to produce 3-HP from renewable sources in cell factories, focusing mainly on *Escherichia coli*, *Klebsiella pneumoniae* and *Saccharomyces cerevisiae*. Despite the significant progress made in this field, commercially exploitable large-scale production of 3-HP in microbial strains has still not been achieved. In this study we investigated the potential of *Bacillus subtilis* as a microbial platform for bioconversion of glycerol into 3-HP. Our recombinant *B. subtilis* strains overexpress the two-step heterologous pathway containing glycerol dehydratase and aldehyde dehydrogenase from *K. pneumoniae*. Genetic engineering, driven by *in silico* optimization, and optimization of cultivation conditions resulted in a 3-HP titer of 10 g/L, in a standard batch cultivation. Our findings provide the first report of successful introduction of the biosynthetic pathway for conversion of glycerol into 3-HP in *B. subtilis*. With this relatively high titer in batch, and the robustness of *B. subtilis* in high density fermentation conditions, we expect that our production strains may constitute a solid basis for commercial production of 3-HP.

Elementary Mode Analysis of the Effect of Redox Balance over H2 Production in Rhodopseudomonas Palustris

- Conference: Synthetic Biology Engineering Evolution Design SEED
- Year: 2017
- Proceeding: 2017 Synthetic Biology: Engineering, Evolution & Design (SEED)
- Group: Poster Session
- Session:

  Confirmed Posters

Authors:
Saldivar-García, A., Instituto Politécnico Nacional
Gonzalez-Garcia, R. A., The University of Queensland
Salgado Manjarrez, E., Unidad Profesional Interdisciplinaria de Biotecnología, Instituto Politecnico Nacional
García-Peña, E. I., Instituto Politécnico Nacional
The adoptive transfer of T cells expressing chimeric antigen receptors (CARs) has demonstrated clinical efficacy in the treatment of advanced cancers, with anti-CD19 CAR-T cells achieving up to 90% complete remission among patients with relapsed B-cell malignancies. However, challenges such as antigen escape and immunosuppression limit the long-term efficacy of adoptive T-cell therapy. Here, I will discuss the development of next-generation T cells that can target multiple cancer antigens and resist immunosuppression, thereby increasing the robustness of therapeutic T cells against tumor defense mechanisms. This presentation will highlight the potential of synthetic biology in generating novel mammalian cell systems with multifunctional outputs for therapeutic applications.

Engineered Control Systems for Asymmetric Cell Division in Synthetic Bacteria

- Conference: Synthetic Biology Engineering Evolution Design SEED
- Year: 2017
- Proceeding: 2017 Synthetic Biology: Engineering, Evolution & Design (SEED)
- Group: Poster Session
- Session:

  Confirmed Posters

Authors:
Molinari, S., Rice University
Shis, D. L., Rice University
Igoshin, O., Rice University
Bennett, M. R., Rice University

Synthetic Biology aims to implement new cellular features through the design of original genetic parts and systems. A wide variety of circuit architectures have been built successfully, proving the ability of synthetic genetic parts to predictably interact with each other within a single bacterial cell [4]. The next stage of synthetic biology will involve the *de novo* engineering of self-organizing multicellular systems. One of the main methods found in nature for creating complex spatially distributed systems is cellular differentiation through asymmetric cell division (ACD), whereby stem cells spawn off irreversibly differentiated daughters. The presence of different cell lineages within an eukaryotic tissue allows for the partitioning of metabolic and structural tasks. Furthermore, if one or both of the daughter cells becomes incapable of dividing, controlling asymmetric division probability can lead to tissue size and shape control [5]. In complex eukaryotic organisms ACD is typically brought about by the differential expression of key genes within each cell type. Here, we engineered a novel method for generating ACD in *Escherichia coli* and regulating its frequency within the population. This work provides a platform for the construction of complex synthetic tissues and organisms from bacterial cells that differentiate during development of the population.

Current methods for creating differentiated cell types in synthetic bacteria utilize gene regulatory networks that allow for bistable gene expression [2, 1]. While such systems generate bimodal populations, they are very sensitive to both intrinsic and extrinsic noise sources that can randomly force transitions between states [3]. We use a novel method for creating terminally differentiated synthetic bacterial cells. Specifically, we refactored DNA shuttling mechanisms from *Caulobacter crescentus*

into *E. coli* to target plasmid DNA into one and only one daughter cell upon division. Thus, the two daughter cells will not just have different gene expression states, but will be irreversibly genetically distinct. Our findings will lead to synthetic "tissues" that can be used in complex settings.

References

[1] R. G. Egbert and E. Klavins. Fine-tuning gene networks using simple sequence repeats.*Proceedings of the National Academy of Sciences*, 109(42):16817â□“16822, 2012.

[2] T. S. Gardner, C. R. Cantor, and J. J. Collins. Construction of a genetic toggle switch in *Escherichia coli*. *Nature*, 403(6767):339â□“342, 2000.

[3] C. Gupta, J. M. LÃ³pez, W. Ott, K. JosiÄ‡, and M. R. Bennett. Transcriptional delay stabilizes bistable gene networks. *Physical review letters*, 111(5):058104, 2013.

[4] A. S. Khalil and J. J. Collins. Synthetic biology: applications come of age. *Nature Reviews*

Genetics, 11(5):367â□“379, 2010.

[5] A. D. Lander, K. K. Gokoffski, F. Y. Wan, Q. Nie, and A. L. Calof. Cell lineages and the logic of proliferative control. *PLoS Biol*, 7(1):e1000015, 2009.

POSTER # 93 - SARA MOLINARI - RICE UNIVERSITY The Biodesign Lab RICE

ENGINEERED CONTROL SYSTEMS FOR ASYMMETRIC CELL DIVISION IN SYNTHETIC BACTERIA

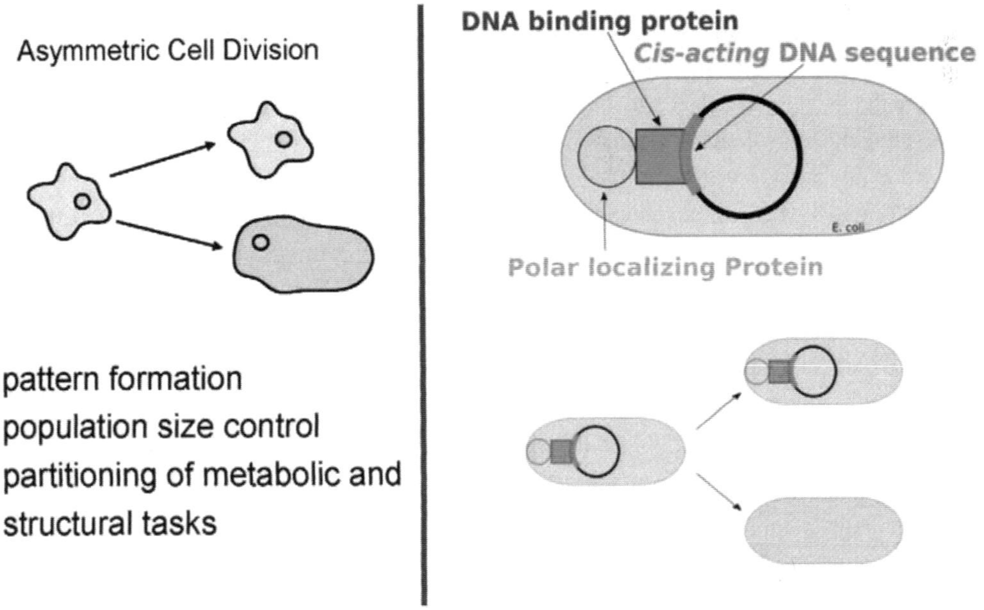

Engineered Protein Materials for Selective Binding of Strategic Metals

- Conference: Synthetic Biology Engineering Evolution Design SEED
- Year: 2017
- Proceeding: 2017 Synthetic Biology: Engineering, Evolution & Design (SEED)
- Group: Poster Session
- Session:

  Confirmed Posters

Authors:
Charrier, M., Joint BioEnergy Institute
Ajo-Franklin, C., Lawrence Berkeley National Laboratory
Borglin, S.

Genetically-encoded biosensors trigger expression of target genes in response to metabolites or small molecules. They offer great potential for diverse applications such as dynamic pathway engineering, environmental sensing, and real-time metabolic monitoring.Â Biosensors can be fine-tuned by modifying their DNA sequences, yet so far we have a poor grasp of how such tuning shapes the resulting dose-response curve. We built and characterized promoter libraries in E. coli controlled by metabolite-responsive transcription factors, and asked how changes in the operator sequence shape the biosensor's dose-response curve. Our data from the lactose and arginine systems revealed that the biosensors' dynamic range and response threshold are inherently coupled to one another. Mathematical modelling revealed a highly nonlinear relation between dynamic range and threshold, exhibiting maximum achievable limits for both parameters. Combining theory and experiments we uncovered two other tuning dials that provide orthogonal control of the dose-response curve. Our work sheds light on the limits of synthetic biology designs and provides guidelines for biosensor design in metabolic engineering.

* Di Liu and Ahmad Mannan contributed equally to this work.

* **Correspondence:** fzhang@seas.wustl.edu, d.oyarzun@imperial.ac.uk

Engineered Proteins Program Mammalian Cells to Target Tnf? Sources

- Conference: Synthetic Biology Engineering Evolution Design SEED
- Year: 2017
- Proceeding: 2017 Synthetic Biology: Engineering, Evolution & Design (SEED)
- Group: Poster Session
- Session:

  Confirmed Posters

Authors:
Qudrat, A., University of Toronto
Truong, K., University of Toronto
Mosabbir, A.

In certain diseases (e.g. atherosclerosis and cancer), there is a local formation of cell masses (e.g. plaques and tumours), a low pH extracellular microenvironment and the secretion of various pro-inflammatory cytokines such as TNFÎ±. The ability to engineer a cell to seek these TNFÎ± sources allows for targeted and local delivery of therapeutic intervention to disease sites. To impart this ability into cells that do not naturally target TNFÎ± sources, here we introduced a system of proteins: an engineered TNFÎ± chimeric receptor (named TNFR1chi), a previously engineered Ca2+-activated RhoA (named CaRQ), VSVG and thymidine kinase. Upon binding of TNFÎ±, TNFR1chi generates a Ca2+ signal. This Ca2+ signal in turn activates CaRQ-mediated non-apoptotic blebs that allow migration towards the TNFÎ± source. Next, with the addition of VSVG, upon low pH induction, these engineered cells fuse with the TNFÎ± source cells. Finally, post-ganciclovir treatment, the cells undergo death via a suicide mechanism assisted by thymidine kinase. Hence, we assembled a system of proteins that forms the basis of engineering a cell to target inflammatory disease sites characterized by the secretion of TNFÎ± and a microenvironment with low pH.

#117

Engineered proteins *program* mammalian cells *to* target TNFα sources

Anam Qudrat
Abdullah Al Mosabbir
Kevin Truong

CIHR IRSC
Institute of Biomaterials & Biomedical Engineering
UNIVERSITY OF TORONTO

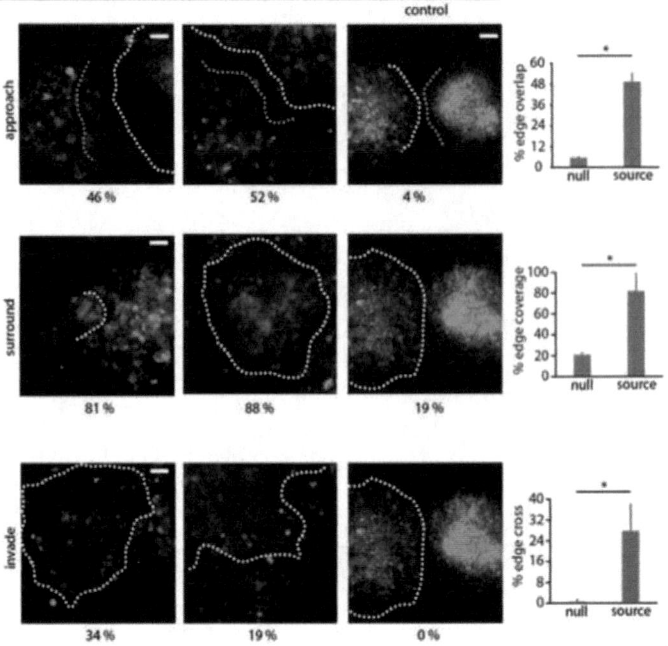

Engineered Strains of the Unicellular Cyanobacterium Synechocystis PCC 6803 Containing Multiple Copies of Phosphoenolpyruvate Carboxylase

- Conference: Synthetic Biology Engineering Evolution Design SEED
- Year: 2017
- Proceeding: 2017 Synthetic Biology: Engineering, Evolution & Design (SEED)
- Group: Poster Session
- Session:

  Confirmed Posters

Authors:
Durall de la Fuente, C., Uppsala University

Improving the RNA switch design process through big data and machine learning

Calvin M. Schmidt, Christina D. Smolke
Stanford University; Stanford, California

RNA switches are a class of functional RNAs that support conditional regulation of gene expression. These genetic switches have broad application in the design of biological systems responsive to changes in their environment with implications for the fields of metabolic engineering, environmental sensing, and therapeutics. The mechanisms underlying the activity of RNA switches rely on programmed structural conformational changes encoded in the sequence of the RNA molecule, implying that an improved elucidation of relationships between sequence, structure, and function can be used to improve our capacities for de novo design. One type of RNA switch is the ribozyme switch, in which binding of the target ligand changes the structure and thus cleavage activity of the switch. Here, we leverage sequencing data and machine learning methods to relate the sequence and structure of RNA molecules to their function and develop models that predict the function of novel ribozyme switches. Through prior work, we have access to data on the activity of hundreds of thousands of ribozyme switch sequences. Using automated structural analysis and machine learning methods, we leveraged a wealth of sequence-activity data on hundreds of thousands of ribozyme switches to develop mathematical models that predict the in vivo activity of a given ribozyme based on its sequence. We have used these models for the de novo design of novel ribozyme switches that exhibit changes in in vivo gene-regulatory activity upon introduction of a target ligand. Our work demonstrates that computational tools can be used to convert RNA sequences into quantitative features that represent key sequence and structural motifs and that these features can be used to develop models that can learn how interactions between these motifs translates to the activity of the functional RNA molecule. The predictive models we developed provide a broadly accessible tool for designing custom biosensors for use in engineering novel biological systems

Rational design of tertiary ribozyme switches through big data and machine learning

Calvin Schmidt
Stanford University
Poster 128

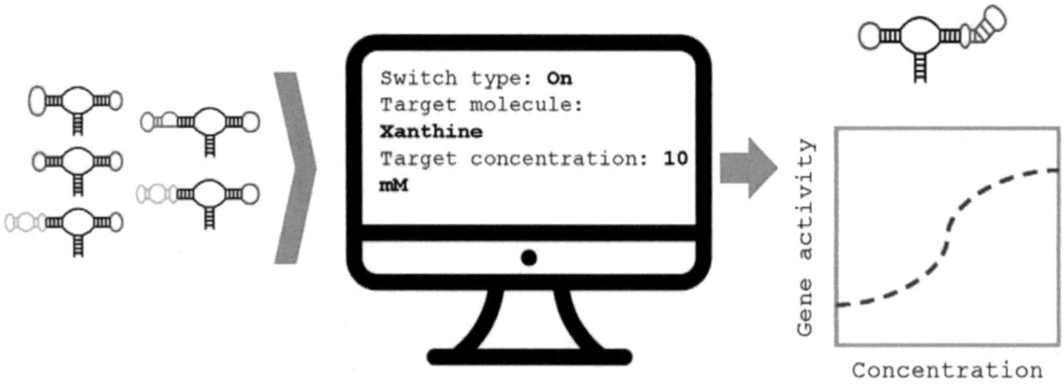

Engineering Autoregulation in Enzymatic Degradation Based Systems for Robust Dynamics and Improved Host Capacity

- Conference: Synthetic Biology Engineering Evolution Design SEED
- Year: 2017
- Proceeding: 2017 Synthetic Biology: Engineering, Evolution & Design (SEED)
- Group: Poster Session
- Session:

  Confirmed Posters

Authors:
Tomazou, M., Imperial College London
Stan, G. B., Imperial College London
Protein glycosylation, the post-translational attachment of complex oligosaccharides (glycans), is the most abundant polypeptide modification in eukaryotes, and is integrally involved in human biology and disease. Changes in glycosylation have been shown to modulate the pharmacokinetics and potency of protein therapeutics and vaccines, making them critical quality attributes for biotechnology applications. However, the study of glycans in native systems (glycobiology) and the intentional manipulation of protein glycosylation patterns (glycoengineering) remain limited by a shortage of methods to characterize glycosylation enzyme activities and a reliance on endogenous glycosylation systems which can produce highly variable glycosylation patterns and constrain choices of production hosts, protein trafficking, and possible glycoforms. The structural complexity of enzymes involved in initial modification of polypeptides with glycans and the vast sequence spaces which must be surveyed to determine their specificities have impeded efforts to implement synthetic glycosylation systems in heterologous production hosts, such as *Escherichia coli*. New tools are required to rapidly express polypeptide modifying glycosyltransferases and determine design rules for engineering protein glycosylation sites which can be efficiently and specifically modified with a glycan of interest.

Here we describe an *in vitro* platform for high-throughput expression and characterization of glycosylation enzymes using *E. coli* Cell-Free Protein Synthesis (CFPS) and Self-Assembled Monolayers for Desorption Ionization Mass Spectrometry (SAMDI-MS). This workflow allowed us to produce >800 µg/mL of a cytoplasmic N-linked glycosyltransferase (NGT) *in vitro* and determine its peptide acceptor and sugar donor specificities at unprecedented depth and throughput with ~10,000 unique reactions conditions and ~3,000 unique peptide substrates. We used this dataset to develop a small, robust acceptor sequence motif (GlycTag) to direct the efficient installation of N-linked glycans onto the internal loops of heterologous protein substrates *in vitro* and in the cytoplasm of living *E. coli*. We found that SAMDI-MS observations at the peptide level accurately predicted modification trends in the context of whole proteins. We placed GlycTags in three structurally diverse target proteins, including the Fc region of a human antibody, and found that these sequences were modified at 3-5 fold greater efficiency compared to wild-type glycosylation sequence motifs. Notably, this N-linked glycosylation system does not require protein transport across cellular membranes or the use of membrane bound components, making it particularly attractive for the development of synthetic glycosylation systems.

This work provides a disruptive technique for glycosyltransferase characterization and provides a new set of tools towards emerging bacterial and *in vitro* glycosylation platforms. Ultimately, these glycoengineering efforts will enable a deeper understanding of glycan structure and function and help to bring about a new generation of rationally designed glycoprotein therapeutics, vaccines, and diagnostics.

Engineering Bacterial S-Layer Protein Arrays to Generate Hybrid Nanomaterials

- Conference: Synthetic Biology Engineering Evolution Design SEED
- Year: 2017
- Proceeding: 2017 Synthetic Biology: Engineering, Evolution & Design (SEED)
- Group: Poster Session
- Session:

  Confirmed Posters

Authors:
Manea, F., Berkeley Lab
Ajo-Franklin, C., Lawrence Berkeley National Laboratory

Synthetic Biology aims to implement new cellular features through the design of original genetic parts and systems. A wide variety of circuit architectures have been built successfully, proving the ability of synthetic genetic parts to predictably interact with each other within a single bacterial cell [4]. The next stage of synthetic biology will involve the *de novo* engineering of self-organizing multicellular systems. One of the main methods found in nature for creating complex spatially distributed systems is cellular differentiation through asymmetric cell division (ACD), whereby stem cells spawn off irreversibly differentiated daughters. The presence of different cell lineages within an eukaryotic tissue allows for the partitioning of metabolic and structural tasks. Furthermore, if one or both of the daughter cells becomes incapable of dividing, controlling asymmetric division probability can lead to tissue size and shape control [5]. In complex eukaryotic organisms ACD is typically brought about by the differential expression of key genes within each cell type. Here, we engineered a novel method for generating ACD in *Escherichia coli* and regulating its frequency within the population. This work provides a platform for the construction of complex synthetic tissues and organisms from bacterial cells that differentiate during development of the population.

Current methods for creating differentiated cell types in synthetic bacteria utilize gene regulatory networks that allow for bistable gene expression [2, 1]. While such systems generate bimodal populations, they are very sensitive to both intrinsic and extrinsic noise sources that can randomly force transitions between states [3]. We use a novel method for creating terminally differentiated synthetic bacterial cells. Specifically, we refactored DNA shuttling mechanisms from *Caulobacter crescentus*

into *E. coli* to target plasmid DNA into one and only one daughter cell upon division. Thus, the two daughter cells will not just have different gene expression states, but will be irreversibly genetically distinct. Our findings will lead to synthetic "tissues" that can be used in complex settings.

References

[1] R. G. Egbert and E. Klavins. Fine-tuning gene networks using simple sequence repeats.*Proceedings of the National Academy of Sciences*, 109(42):16817â□“16822, 2012.
[2] T. S. Gardner, C. R. Cantor, and J. J. Collins. Construction of a genetic toggle switch in *Escherichia coli*. *Nature*, 403(6767):339â□“342, 2000.
[3] C. Gupta, J. M. LÃ³pez, W. Ott, K. JosiÄ‡, and M. R. Bennett. Transcriptional delay stabilizes bistable gene networks. *Physical review letters*, 111(5):058104, 2013.
[4] A. S. Khalil and J. J. Collins. Synthetic biology: applications come of age. *Nature Reviews*

Genetics, 11(5):367â□“379, 2010.
[5] A. D. Lander, K. K. Gokoffski, F. Y. Wan, Q. Nie, and A. L. Calof. Cell lineages and the logic of proliferative control. *PLoS Biol*, 7(1):e1000015, 2009.

Engineering Insulation between Synthetic Bacterial Signaling Pathways

- Conference: Synthetic Biology Engineering Evolution Design SEED
- Year: 2017
- Proceeding: 2017 Synthetic Biology: Engineering, Evolution & Design (SEED)
- Group: Poster Session
- Session:

  Confirmed Posters

Authors:
McClune, C., MIT
Laub, M., MIT
Voigt, C. A., Massachusetts Institute of Technology

Plants are a rich source of unique scaffolds, including 25% of natural-product-derived drugs. However, the discovery, synthesis, and overall material supply chains for sourcing plant natural products and their derivatives remain ad hoc, biased, and tedious. While microbial biosynthesis presents compelling alternatives to traditional approaches based on extraction from natural plant hosts, many challenges exist in the reconstruction of plant specialized metabolic pathways in microbial hosts. My laboratory has developed approaches to address the challenges that arise in the reconstruction of complex biosynthesis schemes, including spatial engineering strategies to direct the activities and specificities of pathway enzymes and recoding strategies to address folding, processing, and stability issues that may arise with the expression of plant enzymes in heterologous microbial hosts. We have utilized these strategies to develop yeast-based production platforms for an important class of plant alkaloids, the benzylisoquinoline alkaloids, including the medicinal opioids. These synthetic biology platforms will lead to transformative advances in natural product discovery, drug development, and production.

Engineering Microbial Consortia Using Fosmid Pool Selection

- Conference: Synthetic Biology Engineering Evolution Design SEED
- Year: 2017
- Proceeding: 2017 Synthetic Biology: Engineering, Evolution & Design (SEED)
- Group: Poster Session
- Session:

  Confirmed Posters

Authors:
Ievdokymenko, K., University of British Columbia
Hallam, S. J., University of British Columbia

The availability of cheap and scalable gene synthesis techniques will open up the exploration of sequence-function relationships at unprecedented scales. Attempts to reduce the costs of gene synthesis have focused on how to assemble microarray derived oligo (OLS) pools into longer length genes. Here we present a new approach capable of assembling hundreds of genes simultaneously from a complex oligo pool by localizing only the fragments required for each individual gene assembly to barcoded beads before isolating the assembly reaction inside an emulsion droplet. Multiplexed reactions of 384 genes, each assembled from five oligos, have costs approaching $1 per 500 bp gene. We apply this to assemble a thousand phylogenetically diverse orthologs of phosphopantetheine adenylyltransferase (PPAT) representing over 500 kbp of coding sequence and use a sequencing based barcode readout to carry out a complementation assay in E. coli, highlighting how this technique can be applied to determine functional relationships of long sequences in high-throughput at low-cost.

Engineering Microbial Utilization of Pectin-Rich Biomass for Producing Commodity Chemicals

- Conference: Synthetic Biology Engineering Evolution Design SEED
- Year: 2017
- Proceeding: 2017 Synthetic Biology: Engineering, Evolution & Design (SEED)
- Group: Poster Session
- Session:

  Confirmed Posters

Authors:
Protzko, R.
Dueber, J., University of California, Berkeley

The prokaryotic CRISPR/Cas9 system enables precise changes at genomic targets. Due to the ease of its customization, Cas9 promises to make high-throughput genetic modifications possible. However, as Cas9 is a bacterially derived system, it has not evolved to interact with the mammalian genome which is compartmentalized and packaged into a nucleo-protein complex known as chromatin. We investigated the mechanism by which dynamic human chromatin states impede efficient Cas9 editing and identified a potential strategy to improve Cas9 efficiency. Using a transgenic HEK293 cell line, we induced accumulation of a Polycomb-based repressor (Gal4-EED) at a luciferase reporter gene to achieve partially and fully silenced chromatin states. We used a transactivator (Gal4-p65) to stimulate a hyperactive state. At silenced luciferase, we confirmed accumulation of the silencing mark H3K27me3 along luciferase by ChIP-qPCR. We measured Cas9-mediated editing using SURVEYOR assay and Cas9 binding using ChIP-qPCR. Interestingly, of the nine gRNA target sites tested along luciferase, Cas9 binding and cutting were inhibited at just five of the sites; four of the target sites were edited at a similar efficiency in each chromatin state. We also observed that a hyperactive state driven by Gal4-p65 reduces Cas9 editing, suggesting high expression levels could be a barrier to efficient Cas9 targeting. To determine a strategy for improving CRISPR efficiency, we broadly inhibited silenced chromatin with either siRNA against Polycomb or the chromatin disrupting drug UNC1999. Increased luciferase expression, indicating an active open chromatin state, was accompanied by increased Cas9 editing efficiency. Finally, we have designed a new transgenic system (inducible Gal4-chromatin remodeler fusions) to determine whether a DNA-targeted chromatin-remodeling proteins can open chromatin at a single locus for more precise enhancement of CRISPR accessibility. Ongoing work includes mapping of nucleosomes at CRISPR target sites to gain a deeper understanding of Cas9-chromatin interaction in human cells. So far, our work represents a significant advance towards understanding the mechanism and extent of Cas9-inhibition by dynamic chromatin. Artificial chromatin remodeling offers a promising solution for efficient binding and editing at sites located in silenced chromatin.

Engineering of Lactobacillus Reuteri As a Biotherapeutic Delivery System

- Conference: Synthetic Biology Engineering Evolution Design SEED
- Year: 2017
- Proceeding: 2017 Synthetic Biology: Engineering, Evolution & Design (SEED)
- Group: Poster Session
- Session:

  Confirmed Posters

Authors:
Ortiz-Vélez, L.
Schaefer, L.
Shulman, A.
Tabor, J.
Britton, R.

We have engineered *Escherichia coli* to overproduce aliphatic methyl ketones (MK) in the C_{11} to C_{15} (diesel) range; this group of MK includes 2-undecanone and 2-tridecanone, which have favorable cetane numbers and are also of importance to the flavor and fragrance industry. Overall, we have made specific improvements that resulted in 50,000-fold enhancement in MK titer relative to that of a fatty acid-overproducing *E. coli* strain. The first generation of engineered *E. coli* (Goh et al. 2012) had modifications that included: (a) overproduction of Î²-ketoacyl-coenzyme A (CoA) thioesters achieved by modification of the Î²-oxidation pathway (specifically, overexpression of a heterologous acyl-CoA oxidase and native FadB, and chromosomal deletion of *fadA*) and (b) overexpression of a native thioesterase (FadM). Subsequent genetic modifications included balancing overexpression of *fadR* and *fadD* to increase fatty acid flux into the pathway, consolidation of the pathway from two plasmids into one, codon optimization, and knocking out key acetate production pathways (resulting in strain EGS1895). This strain produced an MK titer of 1.4 g/L with 1% glucose in shake-flask experiments, representing 40% of the maximum theoretical yield. ^{13}C-based metabolic flux analysis with EGS1895 showed increased carbon flux through the pentose phosphate pathway (PPP), which we hypothesized is to generate more NADPH - an important co-factor for FabG (Î²-ketoacyl-ACP reductase) activity during fatty acid biosynthesis.Â In an attempt to divert flux away from PPP, we overexpressed alternate FabG variants that we have demonstrated preferentially use NADH instead of NADPH. Specifically, overexpression of the NADH-dependent *Acholeplasma laidlawii* FabG increased MK production by 20-50% in shake-flask experiments and had more sustained MK production in fed-batch fermentation compared to EGS1895. We have achieved MK titers of ~6.0 g/L in fed-batch fermentation, which is the best MK titer observed to date.

Engineering Pulsatile Communication in Bacterial Consortia

- Conference: Synthetic Biology Engineering Evolution Design SEED
- Year: 2017
- Proceeding: 2017 Synthetic Biology: Engineering, Evolution & Design (SEED)
- Group: Poster Session
- Session:

  Confirmed Posters

Authors:
Parkin, J.
Hsiao, V., Amyris
Murray, R. M., California Institute of Technology
Protein glycosylation, the post-translational attachment of complex oligosaccharides (glycans), is the most abundant polypeptide modification in eukaryotes, and is integrally involved in human biology and disease. Changes in glycosylation have been shown to modulate the pharmacokinetics and potency of protein therapeutics and vaccines, making them critical quality attributes for biotechnology applications. However, the study of glycans in native systems (glycobiology) and the intentional manipulation of protein glycosylation patterns (glycoengineering) remain limited by a shortage of methods to characterize glycosylation enzyme activities and a reliance on endogenous glycosylation systems which can produce highly variable glycosylation patterns and constrain choices of production hosts, protein trafficking, and possible glycoforms. The structural complexity of enzymes involved in initial modification of polypeptides with glycans and the vast sequence spaces which must be surveyed to determine their specificities have impeded efforts to implement synthetic glycosylation systems in heterologous production hosts, such as *Escherichia coli*. New tools are required to rapidly express polypeptide modifying glycosyltransferases and determine design rules for engineering protein glycosylation sites which can be efficiently and specifically modified with a glycan of interest.

Here we describe an *in vitro* platform for high-throughput expression and characterization of glycosylation enzymes using *E. coli* Cell-Free Protein Synthesis (CFPS) and Self-Assembled Monolayers for Desorption Ionization Mass Spectrometry (SAMDI-MS). This workflow allowed us to produce >800 μg/mL of a cytoplasmic N-linked glycosyltransferase (NGT) *in vitro* and determine its peptide acceptor and sugar donor specificities at unprecedented depth and throughput with ~10,000 unique reactions conditions and ~3,000 unique peptide substrates. We used this dataset to develop a small, robust acceptor sequence motif (GlycTag) to direct the efficient installation of N-linked glycans onto the internal loops of heterologous protein substrates *in vitro* and in the cytoplasm of living *E. coli*. We found that SAMDI-MS observations at the peptide level accurately predicted modification trends in the context of whole proteins. We placed GlycTags in three structurally diverse target proteins, including the Fc region of a human antibody, and found that these sequences were modified at 3-5 fold greater efficiency compared to wild-type glycosylation sequence motifs. Notably, this N-linked glycosylation system does not require protein transport across cellular membranes or the use of membrane bound components, making it particularly attractive for the development of synthetic glycosylation systems.

This work provides a disruptive technique for glycosyltransferase characterization and provides a new set of tools towards emerging bacterial and *in vitro* glycosylation platforms. Ultimately, these glycoengineering efforts will enable a deeper understanding of glycan structure and function and help to bring about a new generation of rationally designed glycoprotein therapeutics, vaccines, and diagnostics.

Engineering Synthetic Proteins to Generate Ca2+ Signals in Mammalian Cells

- Conference: Synthetic Biology Engineering Evolution Design SEED
- Year: 2017
- Proceeding: 2017 Synthetic Biology: Engineering, Evolution & Design (SEED)
- Group: Poster Session
- Session:

  Confirmed Posters

Authors:
Qudrat, A., University of Toronto
Truong, K., University of Toronto
3-hydroxypropanoic acid (3-HP) is an important biomass-derivable platform chemical that can be converted into a number of industrially relevant compounds. There have been several attempts to produce 3-HP from renewable sources in cell factories, focusing mainly on *Escherichia coli*, *Klebsiella pneumoniae* and *Saccharomyces cerevisiae*. Despite the significant progress made in this field, commercially exploitable large-scale production of 3-HP in microbial strains has still not been achieved. In this study we investigated the potential of *Bacillus subtilis* as a microbial platform for bioconversion of glycerol into 3-HP. Our recombinant *B. subtilis* strains overexpress the two-step heterologous pathway containing glycerol dehydratase and aldehyde dehydrogenase from *K. pneumoniae*. Genetic engineering, driven by *in silico* optimization, and optimization of cultivation conditions resulted in a 3-HP titer of 10 g/L, in a standard batch cultivation. Our findings provide the first report of successful introduction of the biosynthetic pathway for conversion of glycerol into 3-HP in *B. subtilis*. With this relatively high titer in batch, and the robustness of *B. subtilis* in high density fermentation conditions, we expect that our production strains may constitute a solid basis for commercial production of 3-HP.

Engineering the Cyanobacterium Synechocystis Sp. PCC 6803 Towards Optimized Solutions for Future Biotechnological Applications

- Conference: Synthetic Biology Engineering Evolution Design SEED
- Year: 2017
- Proceeding: 2017 Synthetic Biology: Engineering, Evolution & Design (SEED)
- Group: Poster Session
- Session:

  Confirmed Posters

Authors:
Nagy, C. I., University of Turku
Aro, E. M., University of Turku
Kallio, P.
Patrikainen, P.
Thiel, K., University of Turku
Dandapani, H., University of Turku
Mulaku, E., University of Turku
Kakko, L., University of Turku

Biocatalyst discovery is integral to bioecomony development, enabling design of robust and scalable bioprocesses that are able to compete with the resource intensive petrochemical industry. Uncultivated microbial communities within natural and engineered ecosystems provide a near-infinite reservoir of genomic diversity and metabolic potential that can be harnessed for biocatalyst discovery[1]. To bridge the cultivation gap, functional metagenomic screens have been developed to recover active genes directly from environmental samples[2]. Here, we describe a pipeline for recovery of biomass-deconstructing biocatalysts sourced from pulp and paper mill sludge metagenomes to develop consolidated bioprocesses that convert waste streams into value-added products. High molecular weight DNA was extracted from sludge and used to construct a fosmid library containing 15,000 clones using the copy control system in EPI300â„¢-T1 RÂ *E. coliÂ* based on established robotic methods[3]. Extracted DNA was also used in whole genome shotgun sequencing to compare the metabolic potential of the sludge community with fosmid screening outcomes using MetaPathways[4]Â with specific emphasis on carbohydrate-active enzymes (CAZymes). In total, 32,232 ORFâ□™s hits in CAZy database were obtained including glycoside hydrolases, glycosyl transferases, and carbohydrate binding module families. The fosmid library was screened for glycosidase hydrolase activities using a pool of sensitive fluorogenic glycosides of 6-chloro-4-methylumbelliferoneÂ enabling rapid recovery of Î²-cellobiosidase, mannosidase and xylosidase activitiesÂ 5. A total of 744 clones capable of converting pooled substrates were recovered indicating an extremely high hit rate (1 hit per 43 clones). Downstream characterization including complete fosmid sequencing and subcloning results to identify specific active genes or gene cassettes is currently in progress. These results reinforce the notion that screening libraries sourced from environments naturally enriched for lignocellulosic biomass deconstruction increases screening efficieny and chances of recovering biocatalysts optimized to perform under specific environmental conditions. The combination of robotic library production and pooled screening using enriched DNA sources for lignocellulosic biomass deconstruction provides an example of high throughput experimental design extensible to a wide range of screening targets.

1. Mewis, K., Armstrong, Z., Song, Y. C., Baldwin, S. A., Withers, S. G., & Hallam, S. J. (2013). Biomining active cellulases from a mining bioremediation system. *Journal of biotechnology*, *167*(4), 462-471.
2. Nyyssönen, M., Tran, H. M., Karaoz, U., Weihe, C., Hadi, M. Z., Martiny, J. B., ... & Brodie, E. L. (2013). Coupled high-throughput functional screening and next generation sequencing for identification of plant polymer decomposing enzymes in metagenomic libraries. *Frontiers in microbiology*, *4*, 282.
3. Mewis, K., Taupp, M., & Hallam, S. J. (2011). A high throughput screen for biomining cellulase activity from metagenomic libraries. *JoVE (Journal of Visualized Experiments)*, (48), e2461-e2461.
4. Hanson, N. W., Konwar, K. M., Wu, S. J., & Hallam, S. J. (2014, May). MetaPathways v2. 0: A master-worker model for environmental Pathway/Genome Database construction on grids and clouds. In *Computational Intelligence in Bioinformatics and Computational Biology, 2014 IEEE Conference on*(pp. 1-7). IEEE.
5. Chen, Hong-Ming, et al. "Synthesis and evaluation of a series of 6-chloro-4-methylumbelliferyl glycosides as fluorogenic reagents for screening metagenomic libraries for glycosidase activity." Carbohydrate research 421 (2016): 33-39.

Engineering the Mammalian Gut Microbiome in Situ

- Conference: Synthetic Biology Engineering Evolution Design SEED
- Year: 2017
- Proceeding: 2017 Synthetic Biology: Engineering, Evolution & Design (SEED)
- Group: Poster Session
- Session:

  Confirmed Posters

Authors:
Chen, S., Columbia University
Ronda, C., Columbia University
Cabral, V., Columbia University
Wang, H. H., Columbia University
Protein glycosylation, the post-translational attachment of complex oligosaccharides (glycans), is the most abundant polypeptide modification in eukaryotes, and is integrally involved in human biology and disease. Changes in glycosylation have been shown to modulate the pharmacokinetics and potency of protein therapeutics and vaccines, making them critical quality attributes for biotechnology applications. However, the study of glycans in native systems (glycobiology) and the intentional manipulation of protein glycosylation patterns (glycoengineering) remain limited by a shortage of methods to characterize glycosylation enzyme activities and a reliance on endogenous glycosylation systems which can produce highly variable glycosylation patterns and constrain choices of production hosts, protein trafficking, and possible glycoforms. The structural complexity of enzymes involved in initial modification of polypeptides with glycans and the vast sequence spaces which must be surveyed to determine their specificities have impeded efforts to implement synthetic glycosylation systems in heterologous production hosts, such as *Escherichia coli*. New tools are required to rapidly express polypeptide modifying glycosyltransferases and determine design rules for engineering protein glycosylation sites which can be efficiently and specifically modified with a glycan of interest.

Here we describe an *in vitro* platform for high-throughput expression and characterization of glycosylation enzymes using *E. coli* Cell-Free Protein Synthesis (CFPS) and Self-Assembled Monolayers for Desorption Ionization Mass Spectrometry (SAMDI-MS). This workflow allowed us to produce >800 μg/mL of a cytoplasmic N-linked glycosyltransferase (NGT) *in vitro* and determine its peptide acceptor and sugar donor specificities at unprecedented depth and throughput with ~10,000 unique reactions conditions and ~3,000 unique peptide substrates. We used this dataset to develop a small, robust acceptor sequence motif (GlycTag) to direct the efficient installation of N-linked glycans onto the internal loops of heterologous protein substrates *in vitro* and in the cytoplasm of living *E. coli*. We found that SAMDI-MS observations at the peptide level accurately predicted modification trends in the context of whole proteins. We placed GlycTags in three structurally diverse target proteins, including the Fc region of a human antibody, and found that these sequences were modified at 3-5 fold greater efficiency compared to wild-type glycosylation sequence motifs. Notably, this N-linked glycosylation system does not require protein transport across cellular membranes or the use of membrane bound components, making it particularly attractive for the development of synthetic glycosylation systems.

This work provides a disruptive technique for glycosyltransferase characterization and provides a new set of tools towards emerging bacterial and *in vitro* glycosylation platforms. Ultimately, these glycoengineering efforts will enable a deeper understanding of glycan structure and function and help to bring about a new generation of rationally designed glycoprotein therapeutics, vaccines, and diagnostics.

Engineering Whole Cell Biocatalysts for Consolidated Bioprocessing

- Conference: Synthetic Biology Engineering Evolution Design SEED
- Year: 2017
- Proceeding: 2017 Synthetic Biology: Engineering, Evolution & Design (SEED)
- Group: Poster Session
- Session:

  Confirmed Posters

Authors:
Ievdokymenko, K., University of British Columbia
Av-Shalom, T., University of British Columbia
Goertsen, D., University of British Columbia
Ocampo, J., University of British Columbia
Verstraete, M., University of British Columbia
Hallam, S. J., University of British Columbia
Smit, J. K., University of British Columbia

Heterologously expressingÂ *Shewanella oneidensis*Â MR-1Â cytochromes *c* (cyt *c*) inÂ *Escherichia coli* enables cellular states to be monitored and controlled through electron flow. However, the electron flux in *E. coli* is significantly less than *Shewanella oneidensis* MR-1, thus limiting the extent of our electronic control. We hypothesize that differences in the cytochrome *c* maturation (*ccm*) machinery between these two species may be responsible for the lower expression levels of cyt *c* and extracellular electron transfer (EET) efficiency. In this research, we constructed random mutations within *ccm* genes by prone error PCR and screened for increases in cyt *c* production in *E. coli* in 96-well plates. Two ccmH mutation strains were shown to exhibit increased cyt *c* expression. Since these two mutations occurred in the C-terminal domain of *ccmH*, which has a similar structure of *ccmI* in the *Shewanella oneidensis* MR-1, we also constructed a hybrid *ccmH* comprising the N-terminal domain of *Escherichia coli* and the C-terminal domain of *Shewanella oneidensis* MR-1 ($ccmH^N$:ccmI). Indeed, cells from this engineered $ccmH^N$:ccmI strain possess a redder color and higher expression of cyt *c*. We measured the electron flux out of these modified variants using three-electrode microbial electrochemical experiments and showed that the $ccmH^N$:ccmI strain produces ~40% more current per cell. Thus, modulating maturation of cyt *c* is not only possible, but can improve electron transfer in synthetic biological systems.

Enlarging the Synthetic Biology Toolbox for Pichia Pastoris: Golden Gate Cloning and CRISPR/Cas9

- Conference: Synthetic Biology Engineering Evolution Design SEED
- Year: 2017
- Proceeding: 2017 Synthetic Biology: Engineering, Evolution & Design (SEED)
- Group: Poster Session
- Session:

  Confirmed Posters

Authors:
Gassler, T., Austrian Centre of Industrial Biotechnology (acib)
Prielhofer, R., Department of Biotechnology, BOKU University of Natural Resources and Life Sciences
Heistinger, L.
Zahrl, R.
Zehetbauer, F.
Steiger, M., Department of Biotechnology, BOKU University of Natural Resources and Life Sciences
Sauer, M., Department of Biotechnology, BOKU University of Natural Resources and Life Sciences
Gasser, B.
Mattanovich, D., Department of Biotechnology, BOKU University of Natural Resources and Life Sciences

State-of-the-art strain engineering techniques for the protein producing yeast host *Pichia pastoris* include overexpression of homologous and heterologous genes, and deletion of host genes. For this purpose overexpression vectors and gene deletion methods such as the split marker technique have been established. For metabolic and cell engineering purposes, the simultaneous overexpression of more than one gene is often needed. Previous approaches employing subsequent steps of overexpression and marker recycling were time- and labor-consuming. Therefore, efficient systems allowing multiple gene overexpression are required, that can be stably integrated into the *P. pastoris* genome. To this end, we developed a synthetic biology toolbox based on Golden Gate cloning to enable efficient construction of complex and versatile over-expression vectors. Up to five different expression cassettes, employing a library of promoters and terminators can be combined into one vector, and successfully integrated into the genomic DNA of *P. pastoris* at targeted loci in one step. Recent trends in synthetic biology, however, go into the direction of building up large and complex reaction networks. To allow for clean and unscarred genetic engineering, a CRISPR/Cas9 based method for gene insertions, deletions and replacements was developed, which paves the way for precise genomic rearrangements in *P. pastoris*. By using this technique precise genomic integrations were performed efficiently without integrative selection markers. The repertoire of genetic techniques developed so far, will provide a wide variety of possibilities to engineer *P. pastoris*. Applications for these synthetic biology tools in cell engineering of recombinant *P. pastoris* will be presented.

No. 42

Enlarging the Synthetic Biology Toolbox for *Pichia pastoris*: Golden Gate Cloning and CRISPR/Cas9

Thomas Gassler *et al.*

Why?

- Defined & controlled modifications
- Marker-free integration
- Complex pathways

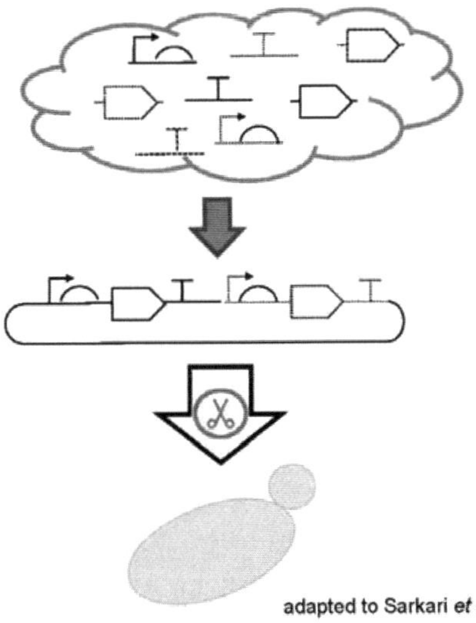

adapted to Sarkari *et al.* (2017)

How?

- Modular Golden Gate cloning
- Efficient CRISPR/Cas9 mediated genome editing

BioToP Biomolecular Technology of Proteins · acib innovations from nature · BOKU · VIBT Vienna Institute of BioTechnology

Austrian Centre of Industrial Biotechnology (acib), Vienna, Austria; BOKU University of Natural Resources and Life Sciences, Vienna, Austria

Establishing a Dual Channel System for Ratiometric Characterisation in Bacillus Subtilis Across Conditions and Gene Contexts

- Conference: Synthetic Biology Engineering Evolution Design SEED
- Year: 2017
- Proceeding: 2017 Synthetic Biology: Engineering, Evolution & Design (SEED)
- Group: Poster Session
- Session:

  Confirmed Posters

Authors:
King, H. J., University of Cambridge

Cell surface glycans make up an important mode of intercellular communication and the enzymes that assemble and degrade them are not only important targets for therapeutics, but also powerful tools for chemical glycobiology. Our recent work has explored 1) the use of engineered glycoside hydrolases for generating universal blood by antigen removal, 2) the identification of inhibitors to glycosyltransferases in cancer pathways, and 3) the development of engineered â□œglycosynthasesâ□□ to assemble and probe the function of glycosaminoglycan structures.

We have previously demonstrated that blood antigen-cleaving glycoside hydrolases can be engineered as tools towards the generation of universal blood through the cleavage of carbohydrate blood group antigens. Identification and directed evolution of these enzymes is enabled by a strategy that involves high-throughput screening with a sensitive fluorescence based assay. Using a similar approach, creatively repurposed to assay for glycosyltransferase rather than glycoside hydrolase activity, we have developed a high-throughput screen for inhibitors targeting fucosyltransferase enzymes that are highly expressed in cancers. Such inhibitors have potential as therapeutics for certain cancers wherein these enzymesâ□™ upregulated activity results in the assembly of carbohydrate structures (e.g. sialyl Lewis^{X}) involved in malignant processes including metastasis. In another focus, we are developing tools for chemoenzymatic glycan synthesis through mechanism-based engineering of â□œglycosynthaseâ□□ enzyme biocatalysts to assemble specific glycosaminoglycan structures. A facile enzymatic synthesis of these glycans will enable glycobiologists to probe their interactions in processes such as neural development and cancer.

Evolution of Chalcone Isomerase from a Non-Enzymatic Ancestor

- Conference: Synthetic Biology Engineering Evolution Design SEED
- Year: 2017
- Proceeding: 2017 Synthetic Biology: Engineering, Evolution & Design (SEED)
- Group: Poster Session
- Session:

  Confirmed Posters

Authors:
Kaltenbach, M., Weizmann Institute of Science
Burke, J. R., The Salk Institute for Biological Sciences
Noel, J. P., The Salk Institute for Biological Studies
Tawfik, D., The Weizmann Institute of Science, Rehovot, Israel
Kamerlin, C. L., Uppsala University
Aharoni, A.

Cell surface glycans make up an important mode of intercellular communication and the enzymes that assemble and degrade them are not only important targets for therapeutics, but also powerful tools for chemical glycobiology. Our recent work has explored 1) the use of engineered glycoside hydrolases for generating universal blood by antigen removal, 2) the identification of inhibitors to glycosyltransferases in cancer pathways, and 3) the development of engineered â□œglycosynthasesâ□□ to assemble and probe the function of glycosaminoglycan structures.

We have previously demonstrated that blood antigen-cleaving glycoside hydrolases can be engineered as tools towards the generation of universal blood through the cleavage of carbohydrate blood group antigens. Identification and directed evolution of these enzymes is enabled by a strategy that involves high-throughput screening with a sensitive fluorescence based assay. Using a similar approach, creatively repurposed to assay for glycosyltransferase rather than glycoside hydrolase activity, we have developed a high-throughput screen for inhibitors targeting fucosyltransferase enzymes that are highly expressed in cancers. Such inhibitors have potential as therapeutics for certain cancers wherein these enzymesâ□™ upregulated activity results in the assembly of carbohydrate structures (e.g. sialyl Lewis^X) involved in malignant processes including metastasis.

In another focus, we are developing tools for chemoenzymatic glycan synthesis through mechanism-based engineering of â□œglycosynthaseâ□□ enzyme biocatalysts to assemble specific glycosaminoglycan structures. A facile enzymatic synthesis of these glycans will enable glycobiologists to probe their interactions in processes such as neural development and cancer.

Experimental Design Considerations for Efficient and Specific Gene Knock-in Using CRISPR-Cas9 for HDR Utilizing Synthetic RNAs

- Conference: Synthetic Biology Engineering Evolution Design SEED
- Year: 2017
- Proceeding: 2017 Synthetic Biology: Engineering, Evolution & Design (SEED)
- Group: Poster Session
- Session:

  Confirmed Posters

Authors:
Schiel, J. A. III, GE Healthcare, Dharmacon
Machado, H., GE Healthcare, Dharmacon
Mayer, M., GE Healthcare, Dharmacon
Chou, E., GE Healthcare, Dharmacon
Basila, M., GE Healthcare, Dharmacon
Kelley, M. L., GE Healthcare, Dharmacon
van Brabant Smith, A., GE Healthcare, Dharmacon

DNA assembly forms the cornerstone of modern synthetic biology. Despite the numerous available methods, scarless multi-fragment assembly of large plasmids remains challenging. Furthermore, the upcoming wave in molecular biological automation demands a rethinking of how we perform DNA assembly. To streamline automation workflow and minimize operator intervention, a non-enzymatic assembly method is highly desirable. Here, we report the optimization and operationalization of a process called Twin-Primer Assembly (TPA), which is a method to assemble PCR-amplified fragments into a plasmid without the use of enzymes.

Each fragment used in TPA requires two PCR products, amplified separately using two sets of primers. One product has an overlap to the fragment that comes before and the other has an overlap to the fragment that comes after. The length of the overlap region is determined by the melting temperature (T_M) of the overlap region, and all overlaps used in an assembly are designed to have the same T_M. The two PCR products are mixed and fully denatured at high temperature before they are allowed to re-anneal slowly, generating intermediates that have overhangs on both sides. These intermediates are then allowed to hybridize at an elevated temperature to form a circular nicked plasmid. Upon *E. coli* transformation, the cells phosphorylate and ligate the nicks to create the final product plasmid.

The TPA non-enzymatic DNA assembly method is scarless and generally sequence independent. It is capable of assembling a 7 kb plasmid from 10 fragments at close to 80% fidelity, and a 31kb plasmid from 5 fragments at around 50% fidelity. Assembling more than 10 fragments is likely possible for constructing smaller plasmids, while assembling a plasmid larger than 31 kb is likely possible using fewer fragments. Even without the use of enzymes, the performance of TPA is on par with some of the best *in vitro* assembly methods currently available. TPA should be an invaluable addition to a synthetic biologistâ□™s toolbox.

Exploiting a Soil Bacterium for the Sustainable Production of High-Value Lipids

- Conference: Synthetic Biology Engineering Evolution Design SEED
- Year: 2017
- Proceeding: 2017 Synthetic Biology: Engineering, Evolution & Design (SEED)
- Group: Poster Session
- Session:

  Confirmed Posters

Authors:
Roccor, R., The University of British Columbia
Eltis, L. D., The University of British Columbia
Round, J., The University of British Columbia

Cell surface glycans make up an important mode of intercellular communication and the enzymes that assemble and degrade them are not only important targets for therapeutics, but also powerful tools for chemical glycobiology. Our recent work has explored 1) the use of engineered glycoside hydrolases for generating universal blood by antigen removal, 2) the identification of inhibitors to glycosyltransferases in cancer pathways, and 3) the development of engineered â□œglycosynthasesâ□□ to assemble and probe the function of glycosaminoglycan structures.

We have previously demonstrated that blood antigen-cleaving glycoside hydrolases can be engineered as tools towards the generation of universal blood through the cleavage of carbohydrate blood group antigens. Identification and directed evolution of these enzymes is enabled by a strategy that involves high-throughput screening with a sensitive fluorescence based assay. Using a similar approach, creatively repurposed to assay for glycosyltransferase rather than glycoside hydrolase activity, we have developed a high-throughput screen for inhibitors targeting fucosyltransferase enzymes that are highly expressed in cancers. Such inhibitors have potential as therapeutics for certain cancers wherein these enzymesâ□™ upregulated activity results in the assembly of carbohydrate structures (e.g. sialyl Lewis^{X}) involved in malignant processes including metastasis. In another focus, we are developing tools for chemoenzymatic glycan synthesis through mechanism-based engineering of â□œglycosynthaseâ□□ enzyme biocatalysts to assemble specific glycosaminoglycan structures. A facile enzymatic synthesis of these glycans will enable glycobiologists to probe their interactions in processes such as neural development and cancer.

Exploring the Evolution of Ahl Lactonases

- Conference: Synthetic Biology Engineering Evolution Design SEED
- Year: 2017
- Proceeding: 2017 Synthetic Biology: Engineering, Evolution & Design (SEED)
- Group: Poster Session
- Session:

  Confirmed Posters

Authors:
Williams-Yuen, J.
Tokuriki, N., University of British Columbia

The field of targeted plant genome editing has experienced exciting progress since the production of the first transgenic plants. Recent advances in targeted genome engineering development of zinc finger nucleases (ZFNs), TAL effector nucleases (TALENs), and CRISPR/Cas systemsâ“ allow modification of DNA sequences of interest, once inside living cells.[1] To introduce genome engineering tools into plant cells and plastids, several methods have been established, such as Agrobacterium or polyethylene glycol (PEG) mediated gene delivery, electroporation, biolistic particle gun, and microinjection. However, current methods are limited by host range limitations, low transfection efficiency, toxicity, and in most cases, transfection limited to protoplast cultures where the cell wall is removed, and not to the mature whole plant.[2] We explore the use of high-aspect ratio nanomaterials for efficient through-cell-wall gene transfer into all desired genotypes of any plant species.

Nanomaterials, with distinct and tunable physical properties (size, shape, and surface charge), effectively deliver biological cargoes to traditionally hard-to-access intracellular locations, such as walled plant cells. We have previously shown that certain nanomaterial formulations can penetrate the mature plant cell, and also the chloroplast.[3,4] Here, we present recent results on the passive delivery of nanoparticles, loaded with plasmid DNA, into walled plant cells and protoplasts. In particular, we examine the delivery mediated by single-walled carbon nanotubes (SWCNT) functionalized with GFP encoding plasmids, and the resulting GFP expression in the transfected cells of mature arugula plants. Our work provides a promising tool for generic, targeted, and passive transport of biomolecules (genes, proteins, etc.) into mature plant cells for targeted genome editing and plastid engineering.

References

1. Li, J.-F. et al. Multiplex and homologous recombination-mediated genome editing in Arabidopsis and Nicotiana benthamiana using guide RNA and Cas9. Nature biotechnology 31, 688-691 (2013).
2. Potrykus, I. Gene Transfer to Plants: Assessment of Published Approaches and Results. Annual Review of Plant Physiology and Plant Molecular Biology 42, 205-225, doi:10.1146/annurev.pp.42.060191.001225 (1991).
3. Giraldo, J. P. et al. Plant nanobionics approach to augment photosynthesis and biochemical sensing. Nat Mater 13, 400-408, doi:10.1038/nmat3890
4. Wong, M. H. et al. Lipid Exchange Envelope Penetration (LEEP) of Nanoparticles for Plant Engineering: A Universal Localization Mechanism. Nano Letters 16, 1161-1172, doi:10.1021/acs.nanolett.5b04467 (2016).

Expression of Heterologous Sigma Factor Expands the Searchable Space for Biofuel Tolerance Mechanisms

- Conference: Synthetic Biology Engineering Evolution Design SEED
- Year: 2017
- Proceeding: 2017 Synthetic Biology: Engineering, Evolution & Design (SEED)
- Group: Poster Session
- Session:

  Confirmed Posters

Authors:
Tomko, T., The University of Vermont
Dunlop, M. J., N/A

Cell surface glycans make up an important mode of intercellular communication and the enzymes that assemble and degrade them are not only important targets for therapeutics, but also powerful tools for chemical glycobiology. Our recent work has explored 1) the use of engineered glycoside hydrolases for generating universal blood by antigen removal, 2) the identification of inhibitors to glycosyltransferases in cancer pathways, and 3) the development of engineered â□œglycosynthasesâ□□ to assemble and probe the function of glycosaminoglycan structures.

We have previously demonstrated that blood antigen-cleaving glycoside hydrolases can be engineered as tools towards the generation of universal blood through the cleavage of carbohydrate blood group antigens. Identification and directed evolution of these enzymes is enabled by a strategy that involves high-throughput screening with a sensitive fluorescence based assay. Using a similar approach, creatively repurposed to assay for glycosyltransferase rather than glycoside hydrolase activity, we have developed a high-throughput screen for inhibitors targeting fucosyltransferase enzymes that are highly expressed in cancers. Such inhibitors have potential as therapeutics for certain cancers wherein these enzymesâ□™ upregulated activity results in the assembly of carbohydrate structures (e.g. sialyl Lewis^{X}) involved in malignant processes including metastasis. In another focus, we are developing tools for chemoenzymatic glycan synthesis through mechanism-based engineering of â□œglycosynthaseâ□□ enzyme biocatalysts to assemble specific glycosaminoglycan structures. A facile enzymatic synthesis of these glycans will enable glycobiologists to probe their interactions in processes such as neural development and cancer.

Extracting input-output relationship of mammalian gene circuits using transient transfections

- Conference: Synthetic Biology Engineering Evolution Design SEED
- Year: 2017
- Proceeding: 2017 Synthetic Biology: Engineering, Evolution & Design (SEED)
- Group: Poster Session
- Session:

  Confirmed Posters

Authors:
Stelzer, C.
Benenson, Y., Swiss Federal Institute of Technology (ETH Zürich)

Bacteria lack compartmentalization and therefore rely on subcellular organization to temporally and spatially localize major cellular processes. As one of the major structures in the cell, the nucleoid possesses positional cues that help direct important functions such as DNA replication and cell division. While some of the factors that contribute to maintaining subcellular organization are characterized, many of the underlying determinants are still unknown. In *Bacillus subtilis*, there remains a class of putative DNA-binding proteins currently uncharacterized, and I hypothesize that some of these proteins use DNA to localize functions important in maintaining subcellular organization. Since many uncharacterized genes lack easily discernable knockout phenotypes, I utilized a gene misexpression approach to identify DNA-binding proteins capable of perturbing some aspect of cellular organization. The misexpression library employed comprises over 800 strains, each containing an uncharacterized gene under the control of an inducible promoter on the *B. subtilis* chromosome. Genes for 98 putative DNA-binding proteins were misexpressed and analyzed for their ability to grow in the presence of inducer. Of these strains, nine exhibited reduced growth and 17 strains did not grow. When analyzed by fluorescence microscopy, 19 of these strains could be associated with a distinct phenotype, including defects in the structure and placement of the chromosome, abnormal cell division, and loss of cell shape. To further characterize strains with the most severe phenotypes we performed RNA-seq experiments to determine if gene misexpression alters the transcription profile. We identified four genes that do not affect transcription, suggesting that the misexpression phenotype is not due to transcriptional regulation. We hypothesize that this class of DNA-binding proteins act as positional cues that help maintain subcellular organization.

Functional Robustness of Synthetic Gene Circuits Under Mutation Pressure

- Conference: Synthetic Biology Engineering Evolution Design SEED
- Year: 2017
- Proceeding: 2017 Synthetic Biology: Engineering, Evolution & Design (SEED)
- Group: Poster Session
- Session:

  Confirmed Posters

Authors:
Lee, J. Y., University of Washington
Kim, K. H., University of Washington

Engineered biological systems hold great potential to transform technologies in health, medicine and sustainability. Challenges in the design of biological devices include our incomplete understanding of the sequence-function relationships underlying the design space, and the limited throughput of experiments required to characterize individual devices. To significantly improve the efficiency of engineering highly functional gene-regulatory RNA devices, we 1) developed a massively parallel FACS-seq screen to assay the activity of hundreds of thousands of ribozyme switch library variants in yeast and mammalian cells, 2) adapted and automated a similarly high-throughput *in vitro* Cleave-seq assay on a liquid handler for parallelization and rapid iteration of the design-build-test cycle, and 3) employed computational analyses to extract design principles from the large datasets to inform future rational design.

To assay their activity in yeast or mammalian cells, ribozyme switches - a class of RNA devices with ligand-controllable self-cleavage activity - are integrated into the untranslated region of a fluorescent reporter, such that cleavage directly modulates fluorescence levels. Combining fluorescence-activated cell sorting (FACS) and next-generation sequencing (NGS), we extensively interrogated the effect of tertiary loop I-loop II interaction motifs on ribozyme activity, characterizing 10^5 variants in a single FACS-seq experiment[1]. Cleave-seq measures *in vitro* cleavage activities of the ribozyme switches at a similar throughput using NGS, and automation of this method allowed us to quickly iterate through different library designs, while obtaining insight into differences in conserved sequence motifs between *in vitro* experiments and inside cells. With these high throughput techniques, we generated new RNA switches to a number of small molecule therapeutics and metabolites, which can serve as gene-control devices and live cell genetic sensors. Our analysis of the large datasets also identified a set of general sequence motifs that result in enhanced ribozyme cleavage activity, highlighting the importance of pseudoknot-like base pairs at the base of the ribozyme stem loops to mediate tertiary interactions. These datasets serve as a rich resource to further enable the discovery of novel hammerhead ribozymes with bioinformatics approaches and streamline the design of RNA switches for synthetic biology applications.

[1]Townshend, B., Kennedy, A. B., Xiang, J. S., & Smolke, C. D. (2015). High-throughput cellular RNA device engineering. Nature methods, 12(10), 989-994.

Genetic Tool Development for Nannochloropsis species

- Conference: Synthetic Biology Engineering Evolution Design SEED
- Year: 2017
- Proceeding: 2017 Synthetic Biology: Engineering, Evolution & Design (SEED)
- Group: Poster Session
- Session:

  Confirmed Posters

Authors:
Ruffing, A., Sandia National Laboratories
McEwen, J.
Strickland, L.
Amyris is an end-to-end biological manufacturer that uses synthetic biology to engineer microbes as catalysts for conversion of sugar into chemical products. Many of these chemical targets were previously derived from limited natural resources and economically inaccessible via traditional chemistries. In the last decade, Amyris has developed a sophisticated technology platform for engineering microbes to produce high levels of isoprenoids including artemisinin and farnesene. In recent years, we have begun to harness the power of this proven infrastructure to expand the portfolio of bio-manufactured chemical products.

In this talk, I will describe the current state-of-the-art tools at Amyris, including our automated strain engineering, high-throughput screening, and fermentation. Then I will discuss how these tools are evolving under the DARPA-funded â□œMgs to Kgsâ□□ project to enable the concurrent development of 100â□™s of products simultaneously. In the first year, the project has produced more than 150 target molecules from all parts of metabolic space. As the project continues, â□œMgs to Kgsâ□□ will enable Amyris to incorporate next generation high-throughput â□˜omics, automation, data science, and machine learning tools to further increase the rate at which biology can be engineered.

Genetically-Encoded Biosensors for Yeast Cell Factory Optimization

- Conference: Synthetic Biology Engineering Evolution Design SEED
- Year: 2017
- Proceeding: 2017 Synthetic Biology: Engineering, Evolution & Design (SEED)
- Group: Poster Session
- Session:

  Confirmed Posters

Authors:
Ambri, F., Technical University of Denmark

Â Francesca Ambri,[1] Michael K. Jensen,[1] Jay D. Keasling[1,2,3,4,5,*]

[1] The Novo Nordisk Foundation Center for Biosustainability, Technical University of Denmark, Denmark
[2] Joint BioEnergy Institute, Emeryville, CA, USA
[3] Biological Systems & Engineering Division, Lawrence Berkeley National Laboratory, Berkeley, CA, USA
[4] Department of Chemical and Biomolecular Engineering & [5] Department of Bioengineering University of California, Berkeley, CA, USA

In the last two decades *S. cerevisiae* has become a commercial workhorse for producing fuels, chemicals, and pharmaceuticals from renewable feedstock (1). However, several factors limit the current adoption of yeast as a chassis for biobased production. One of the main bottlenecks encountered when attempting to engineer yeast for heterologous production of valuable molecules is that only very rarely do such products offer easily identifiable phenotypes that correlate with the productivity of the cell. As such, screening and selection of optimal cell design within a population of engineered variants can become a slow and costly procedure, ultimately hampering the identification of optimal cell designs. To balance the diversity by which genomes can be altered nowadays (1), metabolic engineers are in need of a platform that can couple input (ie. production; metabolite perturbations) with a high-throughput screenable output (ie. GFP; antibiotic resistance). Opposed to eukaryotes, prokaryotes retain an enormous pool of ligand-binding regulators harboring a huge potential reservoir of transcriptional regulators that can be tapped from for the development of biosensors in eukaryotes (2).

Recent studies have shown that performance of transplanted prokaryote transcription factor-based biosensors is largely dependent on the transcriptional level and principally on the operator positioning within the promoter region that lies upstream the reporter gene (3). To confirm and expand these findings we have systematically designed and characterized a test-bed library composed of 30 prokaryote transcriptional regulators for (i) optimal expression level, (ii) common optimal engineering of reporter promoters, and for (iii) enhanced dynamic output ranges. Taken together, our results aim to identify a universally applicable pipeline for optimal transplantation of prokaryotic regulators useful for high-throughput screening in eukaryotes.

1. Zhang, J., Jensen, M. K., & Keasling, J. D. (2015). Development of biosensors and their application in metabolic engineering. *Current Opinion in Chemical Biology*, *28*, 1Ã¢â,¬â□œ8. http://doi.org/10.1016/j.cbpa.2015.05.013
2. Stanton, B. C., Nielsen, A. a K., Tamsir, A., Clancy, K., Peterson, T., & Voigt, C. a. (2014). Genomic mining of prokaryotic repressors for orthogonal logic gates. *Nature Chemical Biology*, *10*(2), 99Ã¢â,¬â□œ105. http://doi.org/10.1038/nchembio.1411
3. Skjoedt, M. L., Snoek, T., Kildegaard, K. R., Arsovska, D., Eichenberger, M., Goedecke, T. J., Keasling, J. D. (2016). Engineering prokaryotic transcriptional activators as metabolite biosensors in yeast, *12*(September). http://doi.org/10.1038/nchembio.2177

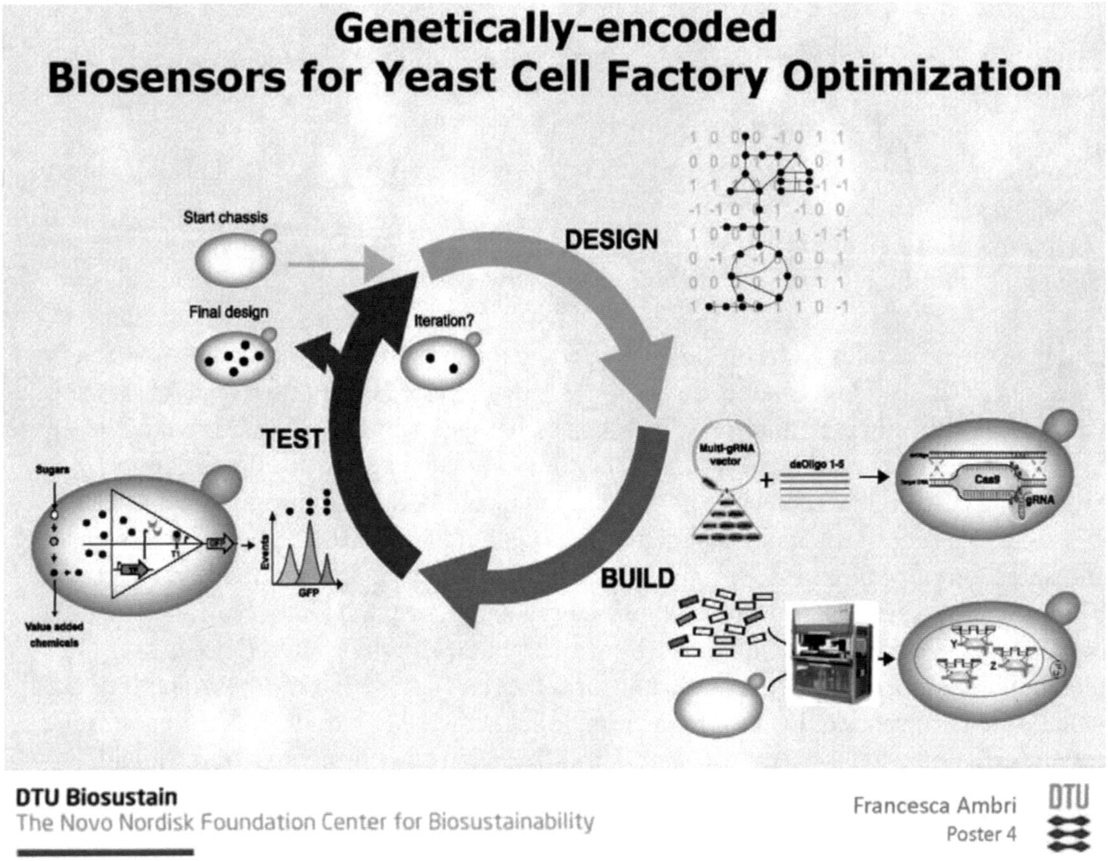

Global Analysis of Promoter Architecture in E. coli

- Conference: Synthetic Biology Engineering Evolution Design SEED
- Year: 2017
- Proceeding: 2017 Synthetic Biology: Engineering, Evolution & Design (SEED)
- Group: Poster Session
- Session:

  Confirmed Posters

Authors:
Urtecho, G.
Insigne, K.
Kosuri, S., University of California, Los Angeles

Countless industries adapt *E. coli* for the production of heterologous resources, such as biofuels, insulin, and other therapeutics. Successfully engineering *E. coli* for these purposes requires the construction of well designed and strictly regulated gene expression networks. In prokaryotes, promoters are the key determinants of gene expression and allow for its flexible regulation. This regulation is mediated by transcription factors that bind specific motifs within the promoter sequence. Variability in the positioning, sequence, and combination of these motifs is extensive and allows for the fine-tuning of downstream gene expression. While this variability allows for precise modulation of gene expression, the resulting complexity makes it difficult to accurately predict the relationship between a promoterâ□™s sequence and its function. Here, we demonstrate a high throughput reporter assay that enables rapid exploration of promoter sequence space. In this assay, synthesized promoters are engineered to express uniquely barcoded transcripts, integrated into a quiescent locus within the *E. coli* genome, and their expression is measured simultaneously using RNA-Seq. Using this approach, we characterize the complexity of endogenous *E. coli* promoters by measuring the expression of over 17,000 loci reported to initiate transcription. We find that a majority of reported transcription start sites (TSSs) are transcriptionally inactive in multiple growth conditions and identify sequence elements responsible for transcriptional activity and differential expression in *E. coli*. Using this approach, we can rapidly test large libraries of designed promoters, thereby improving our understanding of promoter sequence-function relationships and facilitating the development of sequence based models for predicting promoter activity.

Global Analysis of Promoter Architecture in *Escherichia coli*

Guillaume Urtecho & Kimberly Insigne, Hwangbeom Kim, Sriram Kosuri
Department of Chemistry and Biochemistry, University of California at Los Angeles, Los Angeles CA 90095

1)

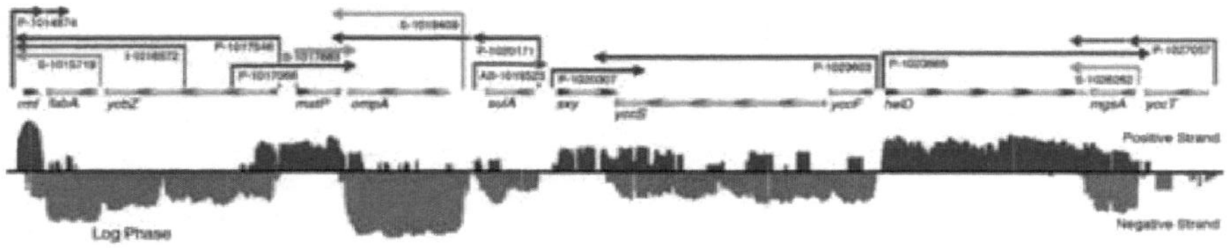

Conway, T. *et al. MBio* 5, 1-12 (2014).

4)

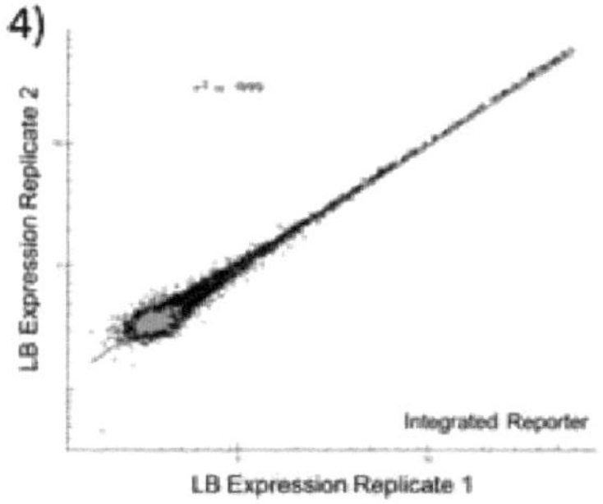

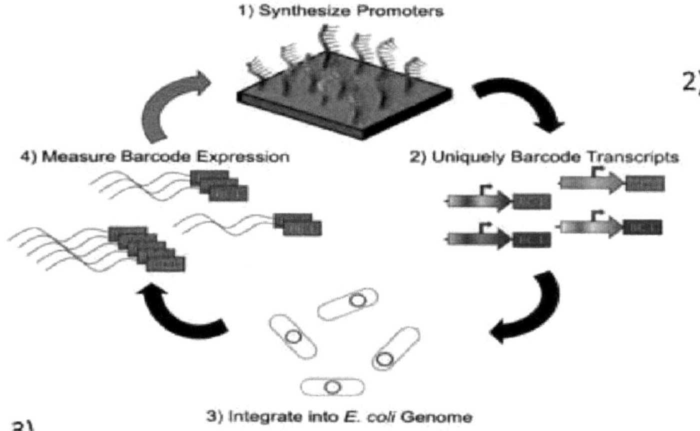

2)

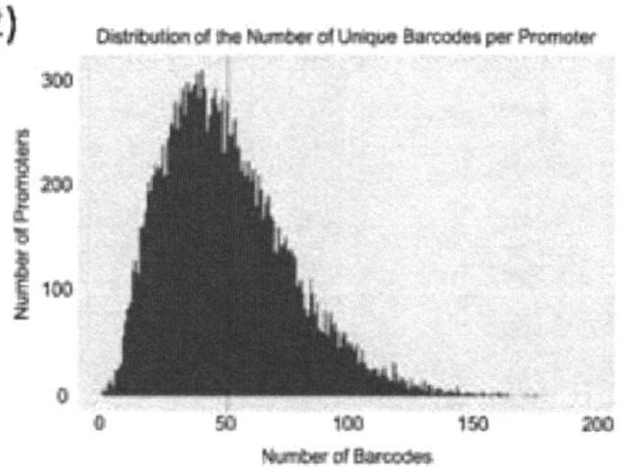

3)

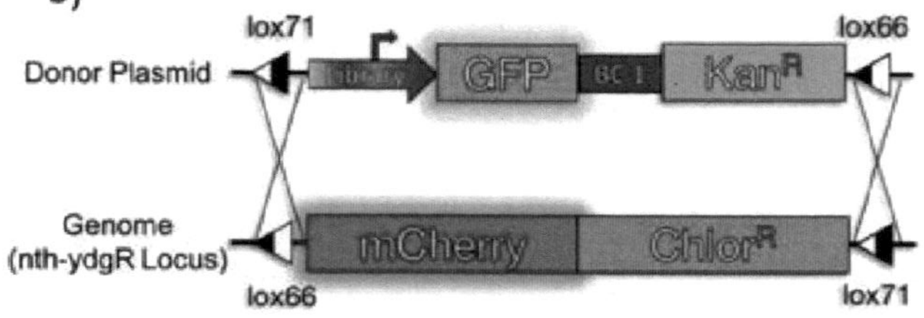

Poster #151, Session A

High Quality, Error-Free DNA for Genome Engineering and Constructing Genetic Circuits for Cell-Free Systems

- Conference: Synthetic Biology Engineering Evolution Design SEED
- Year: 2017
- Proceeding: 2017 Synthetic Biology: Engineering, Evolution & Design (SEED)
- Group: Poster Session
- Session:

  Confirmed Posters

Authors:
Steil, B. P., MYcroarray
Eggenstein, E., MYcroarray
Rouillard, J. M., University of Michigan
Microbes can be engineered to produce biologically-derived replacements for gasoline, diesel, and jet fuel. However, a major technical challenge is that cells can only tolerate limited concentrations of typical fuels. Hydrocarbons are toxic because they damage the cell membrane and increase membrane fluidity. Thus, without further engineering there is a fundamental upper limit to how much biofuel microbes can produce, and cells must trade off production against survival. However, there are many natural examples of microbes that thrive in oil rich environments, such as around natural oil seeps or near oil spills. These microbes harbor mechanisms for tolerating the toxic effects of hydrocarbons. By using library-based approaches, we have identified several novel strategies for improving microbial biofuel tolerance by selecting for genes that enable survival. For example, expression of efflux pumps can dramatically improve tolerance in production strains, and we have further improved survival by designing synthetic feedback systems to control pump expression. This talk will discuss our efforts to enhance microbial synthesis of next-generation biofuels by developing tools for improving microbial tolerance of biofuel production conditions.

High-Throughput Microscopy of E. coli Cells with Micropatterned Agarose Devices

- Conference: Synthetic Biology Engineering Evolution Design SEED
- Year: 2017
- Proceeding: 2017 Synthetic Biology: Engineering, Evolution & Design (SEED)
- Group: Poster Session
- Session:

  Confirmed Posters

Authors:
Priest, D.
Taniguchi, Y.
Tanaka, N.
Our ability to predict protein expression from DNA sequence alone remains poor, reflecting our limited understanding of cis-regulatory grammar and hampering the design of engineered genes for synthetic biology applications. We have generated a model that predicts the translational efficiency of the 5â□™ untranslated region (UTR) of mRNAs in the yeast *Saccharomyces cerevisiae*. We constructed a library of nearly half a million 50 nucleotide-long random 5â□™ UTRs and assayed them in a single massively parallel growth selection experiment. The resulting data have allowed us to quantify the impact on translation of Kozak sequence composition, upstream open reading frames (uORFs) and secondary structure. With this data, we have trained a convolutional neural network on the random library and validated it by predicting the translational efficiency of 5â□™ UTRs that natively occur in yeast. The model additionally was used to computationally evolve highly translating 5â□™ UTRs. We have confirmed experimentally that the great majority of the evolved sequences lead to higher translation rates than the starting sequences, demonstrating the predictive utility of this model.

High-Throughput Mutagenesis of Multiple Components of the Yeast Mating Pathway

- Conference: Synthetic Biology Engineering Evolution Design SEED
- Year: 2017
- Proceeding: 2017 Synthetic Biology: Engineering, Evolution & Design (SEED)
- Group: Poster Session
- Session:

  Confirmed Posters

Authors:
Zimmerman, S.
Dorrity, M. W., University of Washington
Cuperus, J., University of Washington
Fields, S., University of Washington
Queitsch, C.
Plants are a rich source of unique scaffolds, including 25% of natural-product-derived drugs. However, the discovery, synthesis, and overall material supply chains for sourcing plant natural products and their derivatives remain ad hoc, biased, and tedious. While microbial biosynthesis presents compelling alternatives to traditional approaches based on extraction from natural plant hosts, many challenges exist in the reconstruction of plant specialized metabolic pathways in microbial hosts. My laboratory has developed approaches to address the challenges that arise in the reconstruction of complex biosynthesis schemes, including spatial engineering strategies to direct the activities and specificities of pathway enzymes and recoding strategies to address folding, processing, and stability issues that may arise with the expression of plant enzymes in heterologous microbial hosts. We have utilized these strategies to develop yeast-based production platforms for an important class of plant alkaloids, the benzylisoquinoline alkaloids, including the medicinal opioids. These synthetic biology platforms will lead to transformative advances in natural product discovery, drug development, and production.

Immuno-Reprograming of the Tumor Microenvironment Using Synthetic Biotics

- Conference: Synthetic Biology Engineering Evolution Design SEED
- Year: 2017
- Proceeding: 2017 Synthetic Biology: Engineering, Evolution & Design (SEED)
- Group: Poster Session
- Session:

  Confirmed Posters

Authors:
Fisher, A., Synlogic Inc.
Isabella, V., Synlogic Inc.
West, K., Synlogic Inc.
Walsh, K., Synlogic Inc.
Castillo, M., Synlogic Inc.
Ha, B., Synlogic Inc.
Knight, A., Synlogic Therapeutics
Millet, Y., Synlogic Inc.
Lora, J., Synlogic Therapeutics
Miller, P., Synlogic Inc.
Anderson, C., Synlogic, Inc.
Plants are a rich source of unique scaffolds, including 25% of natural-product-derived drugs. However, the discovery, synthesis, and overall material supply chains for sourcing plant natural products and their derivatives remain ad hoc, biased, and tedious. While microbial biosynthesis presents compelling alternatives to traditional approaches based on extraction from natural plant hosts, many challenges exist in the reconstruction of plant specialized metabolic pathways in microbial hosts. My laboratory has developed approaches to address the challenges that arise in the reconstruction of complex biosynthesis schemes, including spatial engineering strategies to direct the activities and specificities of pathway enzymes and recoding strategies to address folding, processing, and stability issues that may arise with the expression of plant enzymes in heterologous microbial hosts. We have utilized these strategies to develop yeast-based production platforms for an important class of plant alkaloids, the benzylisoquinoline alkaloids, including the medicinal opioids. These synthetic biology platforms will lead to transformative advances in natural product discovery, drug development, and production.

Improving Electron Transfer in Cytochrome C-Expressing Escherichia coli by Modulating Cytochrome C Maturation

- Conference: Synthetic Biology Engineering Evolution Design SEED
- Year: 2017
- Proceeding: 2017 Synthetic Biology: Engineering, Evolution & Design (SEED)
- Group: Poster Session
- Session:

  Confirmed Posters

Authors:
Su, L., Lawrence Berkeley National Laboratory
Ajo-Franklin, C., Lawrence Berkeley National Laboratory
Baruch, M., Lawrence Berkeley National Laboratory
Cornejoa, J., Lawrence Berkeley National Laboratory
Fukushimaa, T., Lawrence Berkeley National Laboratory

Heterologously expressingÂ *Shewanella oneidensis*Â MR-1Â cytochromes *c* (cyt *c*) inÂ *Escherichia coli* enables cellular states to be monitored and controlled through electron flow. However, the electron flux in *E. coli* is significantly less than *Shewanella oneidensis* MR-1, thus limiting the extent of our electronic control. We hypothesize that differences in the cytochrome *c* maturation (*ccm*) machinery between these two species may be responsible for the lower expression levels of cyt *c* and extracellular electron transfer (EET) efficiency. In this research, we constructed random mutations within *ccm* genes by prone error PCR and screened for increases in cyt *c* production in *E. coli* in 96-well plates. Two ccmH mutation strains were shown to exhibit increased cyt *c* expression. Since these two mutations occurred in the C-terminal domain of *ccmH*, which has a similar structure of *ccmI* in the *Shewanella oneidensis* MR-1, we also constructed a hybrid *ccmH* comprising the N-terminal domain of *Escherichia coli* and the C-terminal domain of *Shewanella oneidensis* MR-1 ($ccmH^N$:ccmI). Indeed, cells from this engineered $ccmH^N$:ccmI strain possess a redder color and higher expression of cyt *c*. We measured the electron flux out of these modified variants using three-electrode microbial electrochemical experiments and showed that the $ccmH^N$:ccmI strain produces ~40% more current per cell. Thus, modulating maturation of cyt *c* is not only possible, but can improve electron transfer in synthetic biological systems.

Improving the Efficiency of Microbial Isobutanol Synthesis: Directed Evolution Reveals Improved Enzymes and Optimal Enzyme Ratios

- Conference: Synthetic Biology Engineering Evolution Design SEED
- Year: 2017
- Proceeding: 2017 Synthetic Biology: Engineering, Evolution & Design (SEED)
- Group: Poster Session
- Session:

  Confirmed Posters

Authors:
Ghosh, I.
Landick, R., University of Wisconsin-Madison

Efficient microbial synthesis of the specialty biofuel isobutanol is crucial for bioindustrial production, but efficiency is limited by significant bottlenecks within isobutanol-producing metabolic pathways.[1,2] Maximal isobutanol titers will require anaerobic production, which while demonstrated feasible,[3] has not been fully realized due to these bottlenecks. Harnessing directed evolution, we are optimizing the anaerobic synthesis of isobutanol in *Escherichia coli* and *Zymomonas mobilis*, a bacterial strain capable of high-flux sugar catabolism.

For fermentative production of isobutanol in microbes, a valine precursor: Î±-ketoisovalerate is overproduced from pyruvate using AlsS (Acetolactate synthase), IlvC (Ketol-acid reductoisomerase, NADH dependent) and IlvD (Dihydroxy-acid dehydratase), then converted to isobutanol by Kivd (Ketoisovalerate decarboxylase) and AdhA (Alcohol dehydrogenase, NADH dependent).[3,4] Kinetic predictions and previous reports indicate that Î±-ketoisovalerate is likely to be limiting due to low catalytic rates of IlvC and IlvD.[1] This issue is compounded in *Z. mobilis*, where AlsS has a lower affinity for pyruvate than the competing enzyme Pdc (Pyruvate decarboxylase).[5] We aimed to mitigate these bottlenecks by engineered improved versions of these enzymes and identifying optimal ratios of enzyme expression.

We exploited the growth dependence of fermentation-deficient *E. coli* on near-quantitative isobutanol conversion from glucose to identify genetic perturbations that improve metabolic flux. By combinatorically varying enzyme expression levels in an OptSSeq[6] experiment, we identified ratios of enzyme expression that ameriolated bottlenecks in efficient isobutanol synthesis. Additionally, we identified beneficial amino-acid substitutions within enzymes that likely improved catalytic rates, thus enabling even higher efficiency of isobutanol synthesis. We are characterizing the improved enzymatic variants to determine their impacts, and measuring *in-vivo* levels of intermediate metabolites to confirm that our optimized expression cassettes reduce bottlenecks within the pathway.

Initial experiments confirmed difficulties in producing isobutanol in *Z. mobilis*. Insights from our studies in *E. coli* are driving tunings in expression ratios of enzymes with improved catalytic rates. We are also testing various methods of knocking down Pdc in *Z. mobilis*. Our studies will guide significant improvements in microbial production of isobutanol, including from the industrially relevant *Z. mobilis*.

[1]Generoso, *et al* (2015) Curr op in biot 33:1-7
[2]Akita, *et al* (2015) Appl Microbiol Biotechnol 99:991-999
[3]Bastian, *et al* (2011) Metab engg 13:345-352
[4]Atsumi, *et al* (2008) Nature 451:86-89
[5]Neale *et al* (1987) J Bacteriol 169:1024-1028
[6]Ghosh, *et al* (2016) ACS synth biol 5:1519-1534

Improving the efficiency of microbial isobutanol synthesis:
Directed evolution reveals improved enzymes and optimal enzyme ratios

Indro Ghosh, Robert Landick
University of Wisconsin-Madison, WI, USA

Poster # 43

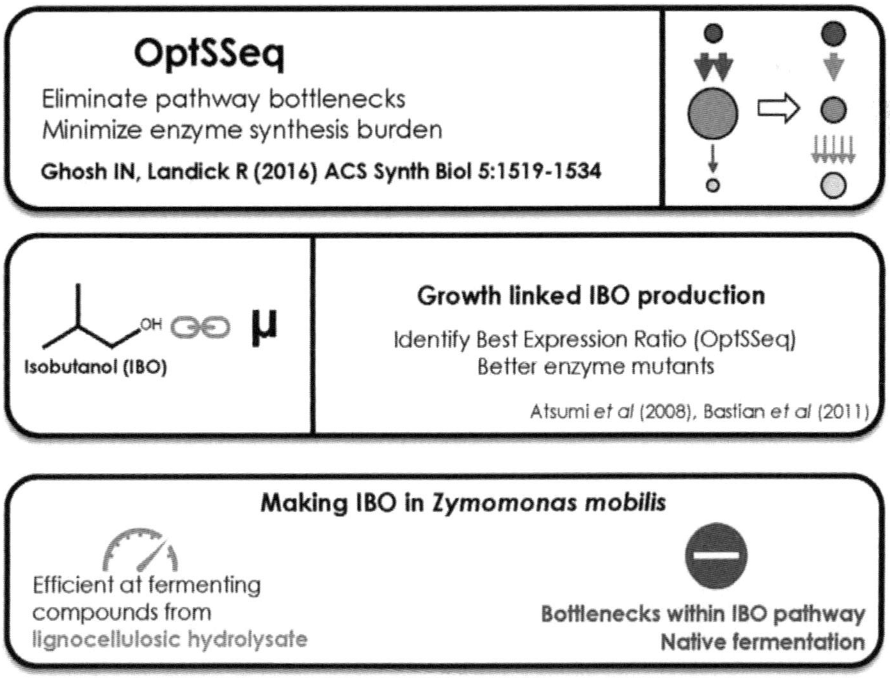

In Vivo Activation of Apo-Hydrogenases Using Synthetic Analogues of Its Active Site in E.coli

- Conference: Synthetic Biology Engineering Evolution Design SEED
- Year: 2017
- Proceeding: 2017 Synthetic Biology: Engineering, Evolution & Design (SEED)
- Group: Poster Session
- Session:

  Confirmed Posters

Authors:
Khanna, N., Uppsala University
Meszaros, L., Uppsala University
Esmieu, C., Uppsala University
Berggren, G., Uppsala University
Lindblad, P., Uppsala University

Cell surface glycans make up an important mode of intercellular communication and the enzymes that assemble and degrade them are not only important targets for therapeutics, but also powerful tools for chemical glycobiology. Our recent work has explored 1) the use of engineered glycoside hydrolases for generating universal blood by antigen removal, 2) the identification of inhibitors to glycosyltransferases in cancer pathways, and 3) the development of engineered â□œglycosynthasesâ□□ to assemble and probe the function of glycosaminoglycan structures.

We have previously demonstrated that blood antigen-cleaving glycoside hydrolases can be engineered as tools towards the generation of universal blood through the cleavage of carbohydrate blood group antigens. Identification and directed evolution of these enzymes is enabled by a strategy that involves high-throughput screening with a sensitive fluorescence based assay. Using a similar approach, creatively repurposed to assay for glycosyltransferase rather than glycoside hydrolase activity, we have developed a high-throughput screen for inhibitors targeting fucosyltransferase enzymes that are highly expressed in cancers. Such inhibitors have potential as therapeutics for certain cancers wherein these enzymesâ□™ upregulated activity results in the assembly of carbohydrate structures (e.g. sialyl Lewis^{X}) involved in malignant processes including metastasis. In another focus, we are developing tools for chemoenzymatic glycan synthesis through mechanism-based engineering of â□œglycosynthaseâ□□ enzyme biocatalysts to assemble specific glycosaminoglycan structures. A facile enzymatic synthesis of these glycans will enable glycobiologists to probe their interactions in processes such as neural development and cancer.

Inducible Genome Modification Utilizing Engineered Ty1 Retrotransposons in Saccharomyces Cerevisiae

- Conference: Synthetic Biology Engineering Evolution Design SEED
- Year: 2017
- Proceeding: 2017 Synthetic Biology: Engineering, Evolution & Design (SEED)
- Group: Poster Session
- Session:

  Confirmed Posters

Authors:
Dorr, B., Arizona State University
Wang, X., Boston University
Standage-Beier, K., Arizona State University

Cell surface glycans make up an important mode of intercellular communication and the enzymes that assemble and degrade them are not only important targets for therapeutics, but also powerful tools for chemical glycobiology. Our recent work has explored 1) the use of engineered glycoside hydrolases for generating universal blood by antigen removal, 2) the identification of inhibitors to glycosyltransferases in cancer pathways, and 3) the development of engineered â□œglycosynthasesâ□□ to assemble and probe the function of glycosaminoglycan structures.

We have previously demonstrated that blood antigen-cleaving glycoside hydrolases can be engineered as tools towards the generation of universal blood through the cleavage of carbohydrate blood group antigens. Identification and directed evolution of these enzymes is enabled by a strategy that involves high-throughput screening with a sensitive fluorescence based assay. Using a similar approach, creatively repurposed to assay for glycosyltransferase rather than glycoside hydrolase activity, we have developed a high-throughput screen for inhibitors targeting fucosyltransferase enzymes that are highly expressed in cancers. Such inhibitors have potential as therapeutics for certain cancers wherein these enzymesâ□™ upregulated activity results in the assembly of carbohydrate structures (e.g. sialyl Lewis^{X}) involved in malignant processes including metastasis. In another focus, we are developing tools for chemoenzymatic glycan synthesis through mechanism-based engineering of â□œglycosynthaseâ□□ enzyme biocatalysts to assemble specific glycosaminoglycan structures. A facile enzymatic synthesis of these glycans will enable glycobiologists to probe their interactions in processes such as neural development and cancer.

Innovative Technology to Produce the Building Blocks for Ambroxide

- Conference: Synthetic Biology Engineering Evolution Design SEED
- Year: 2017
- Proceeding: 2017 Synthetic Biology: Engineering, Evolution & Design (SEED)
- Group: Poster Session
- Session:

  Confirmed Posters

Authors:
Park, G., Evolva

Plants are a rich source of unique scaffolds, including 25% of natural-product-derived drugs. However, the discovery, synthesis, and overall material supply chains for sourcing plant natural products and their derivatives remain ad hoc, biased, and tedious. While microbial biosynthesis presents compelling alternatives to traditional approaches based on extraction from natural plant hosts, many challenges exist in the reconstruction of plant specialized metabolic pathways in microbial hosts. My laboratory has developed approaches to address the challenges that arise in the reconstruction of complex biosynthesis schemes, including spatial engineering strategies to direct the activities and specificities of pathway enzymes and recoding strategies to address folding, processing, and stability issues that may arise with the expression of plant enzymes in heterologous microbial hosts. We have utilized these strategies to develop yeast-based production platforms for an important class of plant alkaloids, the benzylisoquinoline alkaloids, including the medicinal opioids. These synthetic biology platforms will lead to transformative advances in natural product discovery, drug development, and production.

Introducing Catalysis into Non-Catalytic Protein Folds: How Nature Does It

- Conference: Synthetic Biology Engineering Evolution Design SEED
- Year: 2017
- Proceeding: 2017 Synthetic Biology: Engineering, Evolution & Design (SEED)
- Group: Poster Session
- Session:

  Confirmed Posters

Authors:
Kaczmarski, J., Australian National University
Clifton, B., Australian National University
Jackson, C. J., Australian National University
Carr, P., Australian National University
Tokuriki, N., University of British Columbia

The availability of cheap and scalable gene synthesis techniques will open up the exploration of sequence-function relationships at unprecedented scales. Attempts to reduce the costs of gene synthesis have focused on how to assemble microarray derived oligo (OLS) pools into longer length genes. Here we present a new approach capable of assembling hundreds of genes simultaneously from a complex oligo pool by localizing only the fragments required for each individual gene assembly to barcoded beads before isolating the assembly reaction inside an emulsion droplet. Multiplexed reactions of 384 genes, each assembled from five oligos, have costs approaching $1 per 500 bp gene. We apply this to assemble a thousand phylogenetically diverse orthologs of phosphopantetheine adenylyltransferase (PPAT) representing over 500 kbp of coding sequence and use a sequencing based barcode readout to carry out a complementation assay in E. coli, highlighting how this technique can be applied to determine functional relationships of long sequences in high-throughput at low-cost.

Inventory of Composable Elements (ICE): Web of Registries

- Conference: Synthetic Biology Engineering Evolution Design SEED
- Year: 2017
- Proceeding: 2017 Synthetic Biology: Engineering, Evolution & Design (SEED)
- Group: Poster Session
- Session:

  Confirmed Posters

Authors:
Springthorpe, L., Joint Bioenergy Institute
Plahar, H., Joint BioEnergy Institute
Hillson, N. J., Joint BioEnergy Institute
Aravina, E.
Recent studies into the interplay between the microbiome and host have extensively implicated short-chain fatty acid (SCFA) fermentation products as contributors to both intra and extra intestinal health. In the context of the gut, butyrate (a four carbon SCFA) serves as the primary energy source for epithelial colonocytes, drives regulatory T cell differentiation and is therefore essential for the maintenance of colonic mucosal health. On a systemic level, butyrate may influence insulin sensitivity, cholesterol synthesis and autoimmunity. Typically, butyrate is produced by the action of slow-growing Gram positive organisms â□“ often Clostridial in origin â□“ during the fermentation of cellulose and complex carbohydrates. To counteract the depletion of butyrate in the gut, we iteratively engineered *Escherichia coli* Nissle 1917 to produce butyrate. The new butyrate-producing Nissle was capable of millimolar level production of butyrate comparable with the levels produced by enteric Clostridial species. The circuitry was placed under anaerobic control to support induction upon entry into the gastrointestinal track. Finally, proof-of-concept *in vivo* experimentation showed production of butyrate in the mouse gut. This engineered strain may be useful for the treatment of a variety of gastrointestinal and systemic conditions associated with SCFA deficiency.

Leucine Zipper-Mediated Targeting of Multi-Enzyme Cascade Reactions to Inclusion Bodies in Escherichia coli for Enhanced Production of 1-Butanol

- Conference: Synthetic Biology Engineering Evolution Design SEED
- Year: 2017
- Proceeding: 2017 Synthetic Biology: Engineering, Evolution & Design (SEED)
- Group: Poster Session
- Session:

  Confirmed Posters

Authors:
Lee, D. H., Korea Research Institute of Bioscience and Biotechnology
Han, G. H., Korea Research Institute of Bioscience and Biotechnology
Lee, S. G., Korea Research Institute of Bioscience and Biotechnology

Our ability to design cell factories to produce valuable chemicals requires the â□œrecombinationâ□□ not only of existing but also designer enzymes into novel metabolic pathways to achieve entirely new metabolic function. This poses two distinct challenges that need to be solved in an integrated way if we ambition to fully deliver on the promise of synthetic biology. The first challenge deals with the design, at scale, of novel enzymes with high level of activities for reactions not known to be catalyzed in nature. To this end, Arzeda has developed and scaled high-throughput computational methodologies to exploit natural enzyme latent catalytic promiscuity and rapidly design new biocatalysts. A successful industrial application of this technology will be presented.

The second challenge is that of finding optimal ways to arrange natural and designed enzymes to biosynthetize a small-molecule of interest. To illustrate our progress towards solving this challenge, we will discuss a novel computational tool that Arzeda has developed for the automated design of novel biosynthetic pathways. Inspired by retrosynthetic methods in organic chemistry, our software draws on databases of known natural enzymatic reactions as well as reactions that can be catalyzed by designed enzymes to exhaustively enumerate biosynthetic routes from a set of desired metabolites down to any small molecule of interest. Pathways are ranked based on thermodynamic feasibility and other metrics. We will discuss applications of industrial relevance in the field of fine and bulk chemicals.

Mammalian Cells Seek CSF1 Sources Via Engineered Proteins

- Conference: Synthetic Biology Engineering Evolution Design SEED
- Year: 2017
- Proceeding: 2017 Synthetic Biology: Engineering, Evolution & Design (SEED)
- Group: Poster Session
- Session:

  Confirmed Posters

Authors:
Qudrat, A., University of Toronto
Truong, K., University of Toronto
Our ability to predict protein expression from DNA sequence alone remains poor, reflecting our limited understanding of cis-regulatory grammar and hampering the design of engineered genes for synthetic biology applications. We have generated a model that predicts the translational efficiency of the 5â□™ untranslated region (UTR) of mRNAs in the yeast *Saccharomyces cerevisiae.* We constructed a library of nearly half a million 50 nucleotide-long random 5â□™ UTRs and assayed them in a single massively parallel growth selection experiment. The resulting data have allowed us to quantify the impact on translation of Kozak sequence composition, upstream open reading frames (uORFs) and secondary structure. With this data, we have trained a convolutional neural network on the random library and validated it by predicting the translational efficiency of 5â□™ UTRs that natively occur in yeast. The model additionally was used to computationally evolve highly translating 5â□™ UTRs. We have confirmed experimentally that the great majority of the evolved sequences lead to higher translation rates than the starting sequences, demonstrating the predictive utility of this model.

Managing Data with the Experiment Data Depot

- Conference: Synthetic Biology Engineering Evolution Design SEED
- Year: 2017
- Proceeding: 2017 Synthetic Biology: Engineering, Evolution & Design (SEED)
- Group: Poster Session
- Session:

  Confirmed Posters

Authors:
Morrell, W., Sandia National Laboratory
Forrer, M., Sandia National Lab
Birkel, G., Lawrence Berkeley National Laboratory
Hillson, N. J., Joint BioEnergy Institute
Garcia-Martin, H., Lawrence Berkeley National Laboratory
Lopez, T., Joint BioEnergy Institute
Backman, T.
Petzold, C. J., Lawrence Berkeley National Laboratory
Baidoo, E., Joint BioEnergy Institute
Ando, D., Joint BioEnergy Institute
Costello, Z.
Vaino, I., Lawrence Berkeley National Laboratory

Synthetic Biology aims to implement new cellular features through the design of original genetic parts and systems. A wide variety of circuit architectures have been built successfully, proving the ability of synthetic genetic parts to predictably interact with each other within a single bacterial cell [4]. The next stage of synthetic biology will involve the *de novo* engineering of self-organizing multicellular systems. One of the main methods found in nature for creating complex spatially distributed systems is cellular differentiation through asymmetric cell division (ACD), whereby stem cells spawn off irreversibly differentiated daughters. The presence of different cell lineages within an eukaryotic tissue allows for the partitioning of metabolic and structural tasks. Furthermore, if one or both of the daughter cells becomes incapable of dividing, controlling asymmetric division probability can lead to tissue size and shape control [5]. In complex eukaryotic organisms ACD is typically brought about by the differential expression of key genes within each cell type. Here, we engineered a novel method for generating ACD in *Escherichia coli* and regulating its frequency within the population. This work provides a platform for the construction of complex synthetic tissues and organisms from bacterial cells that differentiate during development of the population.

Current methods for creating differentiated cell types in synthetic bacteria utilize gene regulatory networks that allow for bistable gene expression [2, 1]. While such systems generate bimodal populations, they are very sensitive to both intrinsic and extrinsic noise sources that can randomly force transitions between states [3]. We use a novel method for creating terminally differentiated synthetic bacterial cells. Specifically, we refactored DNA shuttling mechanisms from *Caulobacter crescentus*

into *E. coli* to target plasmid DNA into one and only one daughter cell upon division. Thus, the two daughter cells will not just have different gene expression states, but will be irreversibly genetically distinct. Our findings will lead to synthetic "tissues" that can be used in complex settings.

References

[1] R. G. Egbert and E. Klavins. Fine-tuning gene networks using simple sequence repeats.*Proceedings of the National Academy of Sciences*, 109(42):16817â□"16822, 2012.
[2] T. S. Gardner, C. R. Cantor, and J. J. Collins. Construction of a genetic toggle switch in *Escherichia coli*. *Nature*, 403(6767):339â□"342, 2000.
[3] C. Gupta, J. M. LÃ³pez, W. Ott, K. JosiÄ‡, and M. R. Bennett. Transcriptional delay stabilizes bistable gene networks. *Physical review letters*, 111(5):058104, 2013.
[4] A. S. Khalil and J. J. Collins. Synthetic biology: applications come of age. *Nature Reviews*

Genetics, 11(5):367â□"379, 2010.
[5] A. D. Lander, K. K. Gokoffski, F. Y. Wan, Q. Nie, and A. L. Calof. Cell lineages and the logic of proliferative control. *PLoS Biol*, 7(1):e1000015, 2009.

Mapping the Genomic Landscape of CRISPR-Cas9 Cleavage

- Conference: Synthetic Biology Engineering Evolution Design SEED
- Year: 2017
- Proceeding: 2017 Synthetic Biology: Engineering, Evolution & Design (SEED)
- Group: Poster Session
- Session:

  Confirmed Posters

Authors:
Gradia, S., Caribou Biosciences

Engineered biological systems hold great potential to transform technologies in health, medicine and sustainability. Challenges in the design of biological devices include our incomplete understanding of the sequence-function relationships underlying the design space, and the limited throughput of experiments required to characterize individual devices. To significantly improve the efficiency of engineering highly functional gene-regulatory RNA devices, we 1) developed a massively parallel FACS-seq screen to assay the activity of hundreds of thousands of ribozyme switch library variants in yeast and mammalian cells, 2) adapted and automated a similarly high-throughput *in vitro* Cleave-seq assay on a liquid handler for parallelization and rapid iteration of the design-build-test cycle, and 3) employed computational analyses to extract design principles from the large datasets to inform future rational design.

To assay their activity in yeast or mammalian cells, ribozyme switches - a class of RNA devices with ligand-controllable self-cleavage activity - are integrated into the untranslated region of a fluorescent reporter, such that cleavage directly modulates fluorescence levels. Combining fluorescence-activated cell sorting (FACS) and next-generation sequencing (NGS), we extensively interrogated the effect of tertiary loop I-loop II interaction motifs on ribozyme activity, characterizing 10^5 variants in a single FACS-seq experiment[1]. Cleave-seq measures *in vitro* cleavage activities of the ribozyme switches at a similar throughput using NGS, and automation of this method allowed us to quickly iterate through different library designs, while obtaining insight into differences in conserved sequence motifs between *in vitro* experiments and inside cells. With these high throughput techniques, we generated new RNA switches to a number of small molecule therapeutics and metabolites, which can serve as gene-control devices and live cell genetic sensors. Our analysis of the large datasets also identified a set of general sequence motifs that result in enhanced ribozyme cleavage activity, highlighting the importance of pseudoknot-like base pairs at the base of the ribozyme stem loops to mediate tertiary interactions. These datasets serve as a rich resource to further enable the discovery of novel hammerhead ribozymes with bioinformatics approaches and streamline the design of RNA switches for synthetic biology applications.

[1]Townshend, B., Kennedy, A. B., Xiang, J. S., & Smolke, C. D. (2015). High-throughput cellular RNA device engineering. Nature methods, 12(10), 989-994.

Measurements of Translation Initiation from All 64 Codons in E. coli

- Conference: Synthetic Biology Engineering Evolution Design SEED
- Year: 2017
- Proceeding: 2017 Synthetic Biology: Engineering, Evolution & Design (SEED)
- Group: Poster Session
- Session:

  Confirmed Posters

Authors:
Hecht, A., National Institute of Standards and Technology
Glasgow, J., National Institute of Standards and Technology
Jaschke, P., Stanford University
Bawazer, L. A., Stanford University
Munson, M. S., Stanford University
Cochran, J. R., Stanford University
Endy, D., Joint Initiative for Metrology in Biology
Salit, M., Joint Initiative for Metrology in Biology

Our understanding of translation underpins our capacity to engineer living systems. The canonical start codon (AUG) and a few near-cognates (GUG, UUG) are considered as the â□œstart codonsâ□□ for translation initiation in *Escherichia coli*. Translation is typically not thought to initiate from the 61 remaining codons. Here, we quantified translation initiation of green fluorescent protein and nanoluciferase in *E. coli* from all 64 triplet codons and across a range of DNA copy number. We detected initiation of protein synthesis above measurement background for 47 codons. Translation from non-canonical start codons ranged from 0.007% to 3% relative to translation from AUG. Translation from 17 non-AUG codons exceeded the highest reported rates of non-cognate codon recognition. Translation initiation from non-canonical start codons may contribute to the synthesis of peptides in both natural and synthetic biological systems.

The presence of frequent but very low-level expression of proteins via non-canonical start codons would have widespread implications for genome annotation, cellular engineering, and our fundamental understanding of translation initiation. We encourage reconsidering existing definitions and further exploration of what is considered a start codon.

Metabolic Engineering of Pseudomonas Putida S12 for the Production of Styrene

- Conference: Synthetic Biology Engineering Evolution Design SEED
- Year: 2017
- Proceeding: 2017 Synthetic Biology: Engineering, Evolution & Design (SEED)
- Group: Poster Session
- Session:

  Confirmed Posters

Authors:
Otto, M.
Wierckx, N., RWTH Aachen
Blank, L. M., RWTH Aachen University
Kuepper, J.

Rational strain development of microorganisms can enable alternative production processes of styrene, an industrial relevant bulk chemical. However, styrene is toxic and many microorganisms cannot survive in its presence. Recent studies have shown that product toxicity is a major impairment of efficient microbial solvent production with traditional host organisms such as *E. coli* and *S. cerevisiae*.[1,2]Â Bacteria of the genus *Pseudomonas* are known for their extraordinary ability to adapt to challenging environments, making the organism an interesting candidate chassis for styrene production.[3] The aim of this study is the production of styrene via the central metabolite phenylalanine with *Pseudomonas putida* S12. Through the application of targeted genome modifications, the phenylalanine overproducing strain *P. putida* S12 palM12[4]Â was optimized as production host. Codon-optimized genes encoding a phenylalanine ammonia lyase from *A. thaliana* and ferulic acid decarboxylase from *S. cerevisiae* were co-expressed under the control of different expression systems, enabling the conversion of phenylalanine via cinnamic acid to styrene. The goal is the production of super-saturating amounts of styrene which would greatly facilitate product purification. This is only made possible by the high solvent-tolerance of the engineered *P. putida* S12 strains.

[1]McKenna *et al.*, Metab Eng (2011) 13: 544-554
[2]McKenna *et al.*, Microb Cell Fact (2014) 13:123
[3]Ramos *et al.*, FEMS Microbiol Rev (2015) 39(4):555 -566
[4]Nijkamp *et al.*, Appl Microbiol Biotechnol (2005) 69: 170 -177

POSTER 105

METABOLIC ENGINEERING OF SOLVENT TOLERANT *PSEUDOMONAS PUTIDA* S12 FOR STYRENE PRODUCTION

Maike Otto[1], Jannis Küpper[1], Lars M. Blank[1], Nick Wierckx[1]

[1]iAMB – Institute of Applied Microbiology, ABBt – Aachen Biology and Biotechnology, RWTH Aachen University, Worringerweg 1, 52074 Aachen, Germany

1. Introduction

→ Solvent tolerance

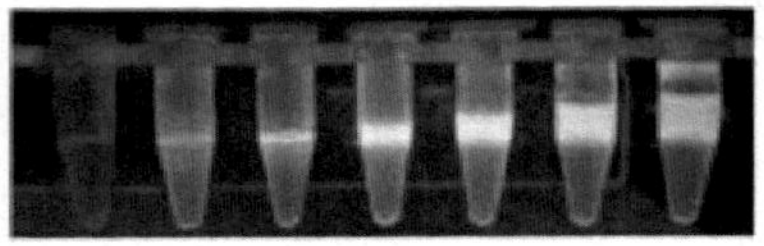

→ Styrene properties

2. Chassis engineering

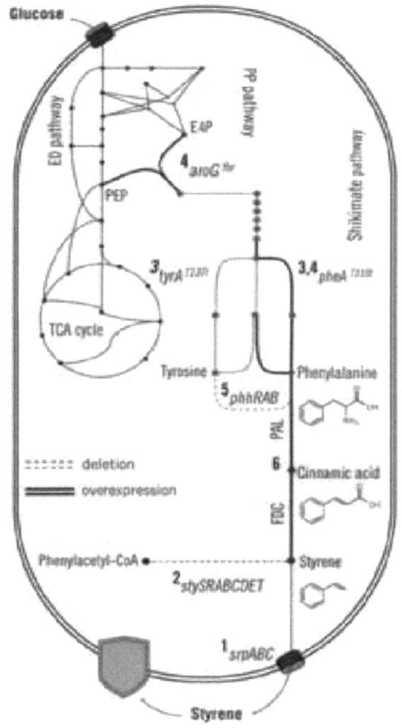

3. Production

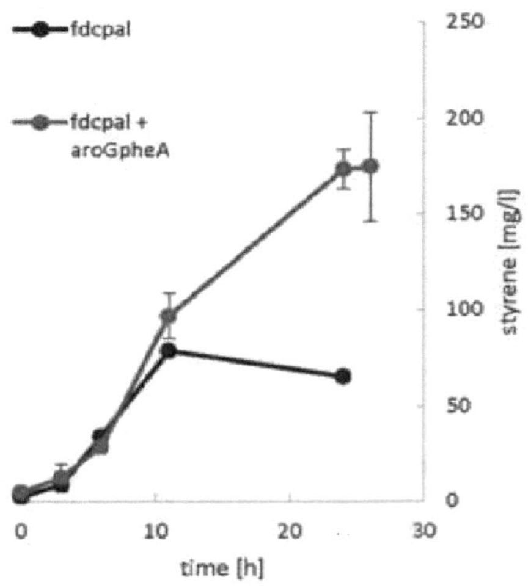

Metagenomic Mining of Natural Prokaryotic Regulatory Elements for Construction of of Trans-Species Genetic Circuits

- Conference: Synthetic Biology Engineering Evolution Design SEED
- Year: 2017
- Proceeding: 2017 Synthetic Biology: Engineering, Evolution & Design (SEED)
- Group: Poster Session
- Session:

  Confirmed Posters

Authors:
Johns, N., Columbia University Medical Center
Wang, H. H., Columbia University
Gomes, A.
CRISPR/Cas9 has wide-reaching scientific and therapeutic applications, but efficient, exact targeting of cutting sites can be challenging. Several cas9 variants have been engineered to target alternate PAMs; however, generation of new variants is often laborious. Additionally, the cutting ability of cas9 nucleases is often highly guide-dependent and shows a large amount of variation between sites. In order to engineer new nuclease variants targeted to cleave a given site, we developed a directed evolution platform to select for DNA cleavage in bacteria. Our system allows rapid generation of mutant libraries, which are challenged with phage containing target sites in a competitive pool. Using this method, we developed *S. aureus* cas9 variants which have altered PAM preferences to target cleavage to novel targets of therapeutic relevance.

Mining the Human Microbiome: Natural Product Discovery from (meta)Genomic Data

- Conference: Synthetic Biology Engineering Evolution Design SEED
- Year: 2017
- Proceeding: 2017 Synthetic Biology: Engineering, Evolution & Design (SEED)
- Group: Poster Session
- Session:

  Confirmed Posters

Authors:
Chang, F. Y., MIT
Clardy, J., Harvard Medical School
Fischbach, M., University of California, San Francisco
Voigt, C. A., Massachusetts Institute of Technology
Mosquito-borne diseases such as Zika, dengue fever and Chikungunya are major and increasing international public health concerns. The primary vector of these viral diseases is *Aedes aegypti* and current control measures against this mosquito are proving inadequate or problematic. An innovative and effective control technique for *Aedes aegypti* involves the release of self-limiting mosquitoes. These mosquitoes are genetically engineered to be homozygous for a self-limiting gene under control of an on/off tetracycline switch. When released into the wild, any offspring of the transgenic males will inherit one copy of the transgene. In the absence of tetracycline antidote, the transgene will prevent the mosquito from developing to adulthood, thus suppressing the population with repeated releases. Oxitecâ□™s self-limiting transgenic insect technology has also been applied successfully to develop population control measures for major crop pests such as the Mediterranean fruit fly (Medfly) and the diamondback moth. In a variation on the self-limiting genetics used in *Aedes* mosquitoes, genetic sexing strains that express the self-limiting gene in a sex-specific manner are used to produce male-only cohorts of these crop pests for release in control programmes. These and other developments in Oxitecâ□™s genetic engineering programme to enable biological control of major insect pests will be discussed. The Oxitec transgenic mosquito method has dramatically reduced *Aedes aegypti* populations in trials around the world and is now being used under operational conditions in the Cayman Islands and Brazil to control *Aedes aegypti*. Results of these programmes will be presented.

Modifying the Cell Surface Carbohydrate Code: From the Evolution of Enzyme Tools for Blood Antigen Removal to the Discovery of Drugs Aimed at Enzyme Targets in Cancer

- Conference: Synthetic Biology Engineering Evolution Design SEED
- Year: 2017
- Proceeding: 2017 Synthetic Biology: Engineering, Evolution & Design (SEED)
- Group: Poster Session
- Session:

  Confirmed Posters

Authors:
Kwan, D. H., Concordia University

Cell surface glycans make up an important mode of intercellular communication and the enzymes that assemble and degrade them are not only important targets for therapeutics, but also powerful tools for chemical glycobiology. Our recent work has explored 1) the use of engineered glycoside hydrolases for generating universal blood by antigen removal, 2) the identification of inhibitors to glycosyltransferases in cancer pathways, and 3) the development of engineered â□œglycosynthasesâ□□ to assemble and probe the function of glycosaminoglycan structures.

We have previously demonstrated that blood antigen-cleaving glycoside hydrolases can be engineered as tools towards the generation of universal blood through the cleavage of carbohydrate blood group antigens. Identification and directed evolution of these enzymes is enabled by a strategy that involves high-throughput screening with a sensitive fluorescence based assay. Using a similar approach, creatively repurposed to assay for glycosyltransferase rather than glycoside hydrolase activity, we have developed a high-throughput screen for inhibitors targeting fucosyltransferase enzymes that are highly expressed in cancers. Such inhibitors have potential as therapeutics for certain cancers wherein these enzymesâ□™ upregulated activity results in the assembly of carbohydrate structures (e.g. sialyl Lewis^X) involved in malignant processes including metastasis. In another focus, we are developing tools for chemoenzymatic glycan synthesis through mechanism-based engineering of â□œglycosynthaseâ□□ enzyme biocatalysts to assemble specific glycosaminoglycan structures. A facile enzymatic synthesis of these glycans will enable glycobiologists to probe their interactions in processes such as neural development and cancer.

Modulating Plant Stress Pathway Output Using Natural Variation

- Conference: Synthetic Biology Engineering Evolution Design SEED
- Year: 2017
- Proceeding: 2017 Synthetic Biology: Engineering, Evolution & Design (SEED)
- Group: Poster Session
- Session:

  Confirmed Posters

Authors:
Cumming, M. N., University of Toronto
Provart, N. J., University of Toronto
A wide variety of amino acyl tRNA synthetases (aaRSs) have had their substrate specificities altered through protein engineering to create mutant aaRSs that can charge tRNA with non-canonical amino acids in cells. While the activities of these mutant aaRSs can be dynamically controlled in cells using conditional promoters, aaRS can also be synthesized using non-natural amino acids as they accumulate in cells, which in some cases could negatively impact their ability to metabolically label other proteins. In addition, our ability to quickly switch aaRS-mediated metabolic labeling on and off remains limited because we have not yet discovered ways to directly control aaRS activity post-translationally in response to environmental changes. To develop prototype aaRSs with activities that are dependent upon direct binding to a chemical, we have engineered a first generation of ligand-responsive aaRSs. For these efforts, we targeted a mutant *Escherichia coli* methionyl tRNA synthetase (MetRS), which was previously developed to charge tRNA with azidonorleucine (Anl) in *Escherichia coli*. We first used a combinatorial approach to discover split MetRS whose fragments cooperatively function when each fragment is fused to a pair of proteins that interact. We then examined whether the activity of split MetRS can be regulated by either fusing the fragments to a pair of proteins whose interaction is stabilized by ligand binding or by fusing the fragments to the termini of a single protein domain that exhibits a ligand-dependent conformational change. We have found that both approaches can be used to regulate Anl labeling of newly synthesized proteins in a ligand-dependent manner. When MetRS fragments were fused to FKBP12 and the rapamycin binding domain of mTOR, metabolic labeling was significantly enhanced in growth medium containing rapamycin, which stabilizes the FKBP12-mTOR complex. In addition, fusion of MetRS fragments to the termini of the ligand-binding domain of the estrogen receptor yielded a protein whose metabolic labeling was significantly enhanced when 4-hydroxytamoxifen was added to the growth medium. Ongoing efforts are focused on assessing whether the ligand-dependent metabolic labeling mediated by our engineered proteins can be switched on faster than labeling achieved using transcriptional regulation, improving the metabolic

labeling efficiencies that can be achieved with our prototype aaRS switches, and investigating the diversity of ligands that can be used to directly regulate MetRS activity. The protein switches arising from these studies are expected to be useful for extending our control over protein metabolic labeling with Anl by allowing for fast regulation of this labeling using post-translational reactions. Furthermore, this approach can be applied to metabolic labeling with additional non-natural amino acids by extending our designs to structurally-related aaRS with distinct substrate specificities. Ligand-responsive aaRSs are expected to have applications in the production of protein biomaterials, proteomic studies, and protein pharmaceutical biosynthesis.

Molecular and Engineering Design Concept of Bacterial Surface Antigen(Fimbriae) Involved in Cancer Therapy

- Conference: Synthetic Biology Engineering Evolution Design SEED
- Year: 2017
- Proceeding: 2017 Synthetic Biology: Engineering, Evolution & Design (SEED)
- Group: Poster Session
- Session:

  Confirmed Posters

Authors:
Brahma, P. N. K. Sr., Institute of genetic Engineering

In evolution of animal kingdom we have observed the paradigm shifts of unicellular organisms to multicellular organism, prokaryotic to eukaryotic cells, simple to complicated immune systems, the genetic mutations and their diversified species description and the cancer generations, the environmental and genetic impacts on species evolution and the debates of Lamack and Darwin's evolutionary theory. In the recent years we have seen the trends is shifting. We have observed the importance of biological evolution has been shifted to immune responses in ensuring the generations of cancer diversified by various organ specific activities and finally we have observed the functions of stem cells in curing cancer as regenerative medicines, where the questions of regenerations and repairing are considered. These repairing mechanisms are confined at the latest studies on CAS9, a RNA based DNA repairing. When the author is justifying the said facts, it was found that all are mainly made in vitro (i.e. on experimental tables) and speculated for target specific activities in vivo. To specify the immune response in vivo, the target specific genetic and protein manipulations were essential, so the application of genetic engineering (GE) in Escherichia coli K-12 was essential and was predominated (i.e the beginning of molecular biology) In this chapter scientist have isolated, transcribed and translated foreign DNA and cloned the said DNA into vector plasmid to design target specific activities (i.e. mainly used in basic studies, gene expressions, hybrid protein based immune response and productions of valuable life saving products) .The author has similarly used the concepts of gene cloning first to study the complicated phenomena of bacterial adherence, immune response in prevention antidiarrheal activities(i.e. to clone donor pathogenic E.coli adherent (colonization Factor Antigen CFA) gene into pBR 322 vector plasmid and to generate hybrid Escherichia coli K-12, which may prevent diarrhea of the donor fatal infection. The author was successful to show the facts of immune response in Balb/c mice through this process of gene cloning and generation of hybrid genetically Engineered (GE) Escherichia coli K-12, was successful. Thereafter the concepts were moved to study the nanochanneling properties of those hybrid fimbriae / pili. In this paper the author will attempt to described the steps involved in such design. The nanochanneling properties of hybrid fimbriae of Escherichia coli K-12 were successfully identified in preparation of hybrid fimbrie' pili based araldite thin film membrane, positive to water channeling, compared to pure araldite. After this design the author shifted his study thoughts / concepts of application towards the BNT (Bionanotube). In which the author proposes that this BNT might be used more efficient compared to CNT (Carbon Nanotube) to prevent side effects (i.e these BNT may be generated from organs, affected by cancer), the isolated gene of the said cells could be used to generate hybrid fimbriae (BNT) and to immobilize cancer drugs additionally. Hybrid GE-BNT will help immune response and drugs will be released for therapeutic applications. Balb/c mice and nanochanneling studies are need to be completed.

Molecular Genetic Characterization of Terreic Acid Pathway in Aspergillus Terreus

- Conference: Synthetic Biology Engineering Evolution Design SEED
- Year: 2017
- Proceeding: 2017 Synthetic Biology: Engineering, Evolution & Design (SEED)
- Group: Poster Session
- Session:

  Confirmed Posters

Authors:
Sun, W.
Guo, C.
Wang, C.
Abstract: Kinetic Design of RNA Architectures for Eengineering and Analyzing Systems on a Large Scale (Synthetic Biology: Engineering, Evolution and Design 2017)

Monitoring Microbial Genetic Circuits in Soils Using Orthogonal Enzymes That Synthesize Volatile Metabolites

- Conference: Synthetic Biology Engineering Evolution Design SEED
- Year: 2017
- Proceeding: 2017 Synthetic Biology: Engineering, Evolution & Design (SEED)
- Group: Poster Session
- Session:

  Confirmed Posters

Authors:
Silberg, J. J., Rice University
Cheng, H. Y., Rice University
Masiello, C. A., Rice University
Del Valle, I., Rice University
Huh, D.
This presentation will focus on Vmaxâ„¢, a high performing bacterium being developed at Synthetic Genomics, Inc. as a next-generation host organism for biotechnology and molecular biology applications[1].Â We will describe the current state of the platform, and will discuss how technology and genome design principles developed over the course of the minimal cell project[2] are being applied to engineer advanced Vmaxâ„¢ strains.

1. Weinstock, M. T., Hesek, E. D., Wilson, C. M. & Gibson, D. G. *Vibrio natriegens* as a fast-growing host for molecular biology. *Nat. Methods* **13,** 849â□"851 (2016).
2. Hutchison, C. A. *et al.* Design and synthesis of a minimal bacterial genome. *Science* **351,** aad6253 (2016).

Multiplex Gene Synthesis in Emulsions

- Conference: Synthetic Biology Engineering Evolution Design SEED
- Year: 2017
- Proceeding: 2017 Synthetic Biology: Engineering, Evolution & Design (SEED)
- Group: Poster Session
- Session:

  Confirmed Posters

Authors:
Plesa, C., University of California, Los Angeles
Kosuri, S., University of California, Los Angeles
Lubock, N. B., University of California, Los Angeles
Sidore, A.
Church, G. M., Harvard Medical School
Zhang, D.

The availability of cheap and scalable gene synthesis techniques will open up the exploration of sequence-function relationships at unprecedented scales. Attempts to reduce the costs of gene synthesis have focused on how to assemble microarray derived oligo (OLS) pools into longer length genes. Here we present a new approach capable of assembling hundreds of genes simultaneously from a complex oligo pool by localizing only the fragments required for each individual gene assembly to barcoded beads before isolating the assembly reaction inside an emulsion droplet. Multiplexed reactions of 384 genes, each assembled from five oligos, have costs approaching $1 per 500 bp gene. We apply this to assemble a thousand phylogenetically diverse orthologs of phosphopantetheine adenylyltransferase (PPAT) representing over 500 kbp of coding sequence and use a sequencing based barcode readout to carry out a complementation assay in E. coli, highlighting how this technique can be applied to determine functional relationships of long sequences in high-throughput at low-cost.

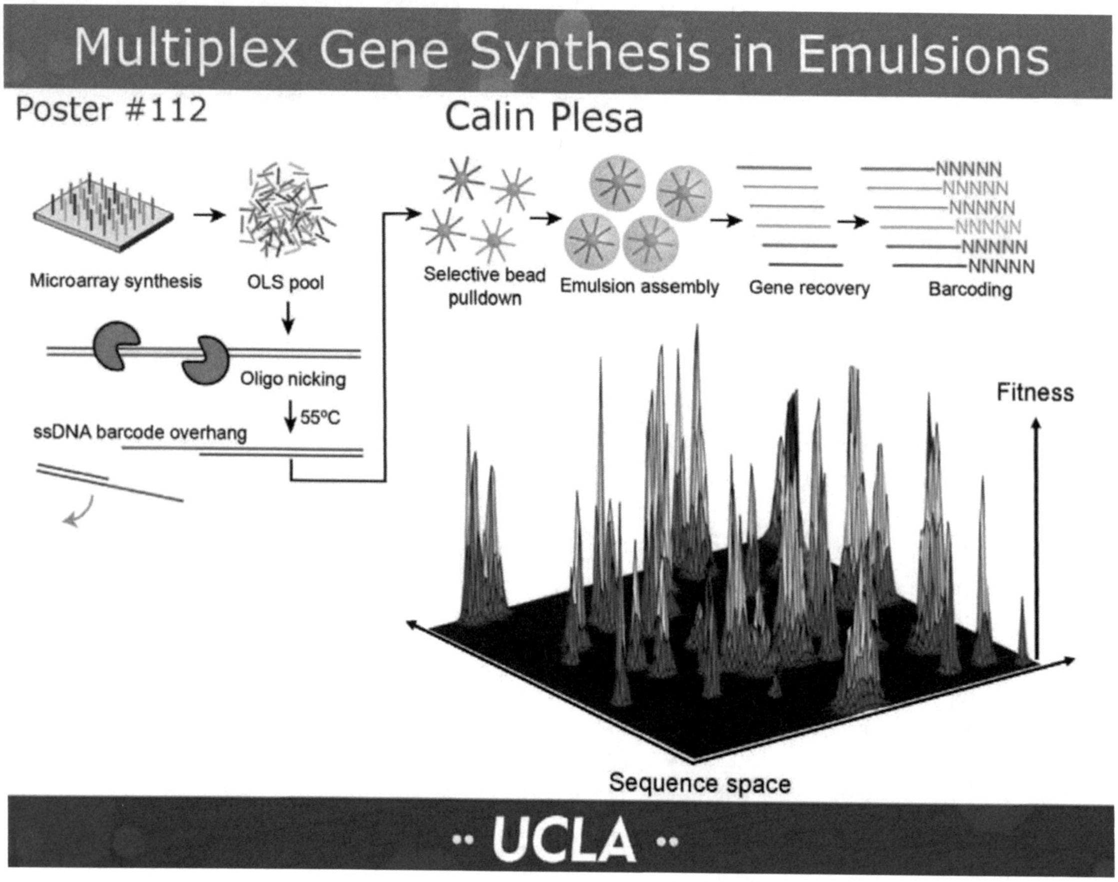
Multiplex Gene Synthesis in Emulsions
Poster #112
Calin Plesa
Microarray synthesis
OLS pool
Oligo nicking
55°C
ssDNA barcode overhang
Selective bead pulldown
Emulsion assembly
Gene recovery
Barcoding
NNNNN
Fitness
Sequence space
UCLA

Naringenin-Responsive Riboswitch-Based Fluorescent Biosensor Module for Escherichia coli Co-Cultures

- Conference: Synthetic Biology Engineering Evolution Design SEED
- Year: 2017
- Proceeding: 2017 Synthetic Biology: Engineering, Evolution & Design (SEED)
- Group: Poster Session
- Session:

  Confirmed Posters

Authors:
Xiu, Y., Rensselaer Polytechnic Institute
Jang, S., POSTECH
Jones, J. A., Rensselaer Polytechnic Institute
Linhardt, R. J., Rensselaer Polytechnic Institute
Yuan, Q., Beijing University of Chemical Technology
Koffas, M. A. G., Rensselaer Polytechnic Institute
Jung, G. Y., POSTECH
Zill, N. A., Rensselaer Polytechnic Institute

Development of advanced functional materials with highly organized nano- and microstructures are of increasing demand across the medical, optical, energy, and mechanical fields. The functionality of materials is dependent on the relationship between structure and composition, demonstrating the importance of controlling molecular assembly throughout a material. In this work, we are applying several of natureâ□™s biological mechanisms for building structurally complex materials to engineer new biomaterials that display a wide range of functional properties. First, we are exploring whether post-translational modifications to the silica mineralizing R5 peptide from diatoms can be used to control silica morphology at the submicron scales. Second, we are growing melanin on R5 to create R5-melanin embedded silica materials. Finally, we are combining R5 with proteins from magnetotactic bacteria to synthesize silica-coated magnetic nanoparticles. By integrating these approaches for controlling material structure and composition, we aim to demonstrate the feasibility of using proteins to synthesize multifunctional materials.

Opportunitis for Facility-Enabled Science at the Doe Joint Genome Institute (JGI)

- Conference: Synthetic Biology Engineering Evolution Design SEED
- Year: 2017
- Proceeding: 2017 Synthetic Biology: Engineering, Evolution & Design (SEED)
- Group: Poster Session
- Session:

  Confirmed Posters

Authors:
Yoshikuni, Y., University of California, Berkeley
Protein glycosylation, the post-translational attachment of complex oligosaccharides (glycans), is the most abundant polypeptide modification in eukaryotes, and is integrally involved in human biology and disease. Changes in glycosylation have been shown to modulate the pharmacokinetics and potency of protein therapeutics and vaccines, making them critical quality attributes for biotechnology applications. However, the study of glycans in native systems (glycobiology) and the intentional manipulation of protein glycosylation patterns (glycoengineering) remain limited by a shortage of methods to characterize glycosylation enzyme activities and a reliance on endogenous glycosylation systems which can produce highly variable glycosylation patterns and constrain choices of production hosts, protein trafficking, and possible glycoforms. The structural complexity of enzymes involved in initial modification of polypeptides with glycans and the vast sequence spaces which must be surveyed to determine their specificities have impeded efforts to implement synthetic glycosylation systems in heterologous production hosts, such as *Escherichia coli*. New tools are required to rapidly express polypeptide modifying glycosyltransferases and determine design rules for engineering protein glycosylation sites which can be efficiently and specifically modified with a glycan of interest.

Here we describe an *in vitro* platform for high-throughput expression and characterization of glycosylation enzymes using *E. coli* Cell-Free Protein Synthesis (CFPS) and Self-Assembled Monolayers for Desorption Ionization Mass Spectrometry (SAMDI-MS). This workflow allowed us to produce >800 μg/mL of a cytoplasmic N-linked glycosyltransferase (NGT) *in vitro* and determine its peptide acceptor and sugar donor specificities at unprecedented depth and throughput with ~10,000 unique reactions conditions and ~3,000 unique peptide substrates. We used this dataset to develop a small, robust acceptor sequence motif (GlycTag) to direct the efficient installation of N-linked glycans onto the internal loops of heterologous protein substrates *in vitro* and in the cytoplasm of living *E. coli*. We found that SAMDI-MS observations at the peptide level accurately predicted modification trends in the context of whole proteins. We placed

GlycTags in three structurally diverse target proteins, including the Fc region of a human antibody, and found that these sequences were modified at 3-5 fold greater efficiency compared to wild-type glycosylation sequence motifs. Notably, this N-linked glycosylation system does not require protein transport across cellular membranes or the use of membrane bound components, making it particularly attractive for the development of synthetic glycosylation systems.

This work provides a disruptive technique for glycosyltransferase characterization and provides a new set of tools towards emerging bacterial and *in vitro* glycosylation platforms. Ultimately, these glycoengineering efforts will enable a deeper understanding of glycan structure and function and help to bring about a new generation of rationally designed glycoprotein therapeutics, vaccines, and diagnostics.

Optimizing the Violacein Biosynthetic Pathway Using Droplet Microfluidics

- Conference: Synthetic Biology Engineering Evolution Design SEED
- Year: 2017
- Proceeding: 2017 Synthetic Biology: Engineering, Evolution & Design (SEED)
- Group: Poster Session
- Session:

  Confirmed Posters

Authors:
Gach, P., Joint BioEnergy Institute
Hillson, N. J., Joint BioEnergy Institute
Adams, P. D., Joint BioEnergy Institute
Singh, A. K., Joint BioEnergy Institute
Keasling, J., University of California, Berkeley
Deutsch, S., Lawrence Berkeley National Laboratory
Nath, S., DOE Joint Genome Institute, Lawrence Berkeley National Laboratory
Kaplan, N., Joint BioEnergy Institute
Raje, M., Joint BioEnergy Institute

Synthetic biology experiments require optimization of pathways consisting of many genes and other genetic elements and given the large number of alternatives available for each element, optimization of a pathway can require large number of experiments. Currently, these experiments are done using fairly large amounts of costly reagents per experiment making the process very expensive, extremely slow and irreproducible. Our lab has previously developed droplet-based microfluidic systems for automating specific procedures of synthetic biology: gene assembly and electroporation [1] or heat-shock and culture [2]. However, these operations were performed separately and required additional manual sample preparation to perform the proof-of-principle experiments.

The current system integrates these processes to provide a programmable device for optimizing biosynthetic pathways. We applied this platform to screen for the production of violacein, a molecule with applications in textiles, agriculture and found to have antibacterial, antiviral and anticancer behaviors [3]. Violacein can be biosynthesized in *E. coli* from L-tryptophan in a process using five enzymes [4]. Digital microfluidics was initially used for dispensing and mixing the gene variants, vector backbone and assembly reagents because of the technologies great control of individual droplets and programmability. This component of the device is capable of generating a combinatorial library composed of 25 variants of the 11 kB violacein producing plasmid VioABCDE. Following combinatorial mixing of gene parts the droplets were transferred to a flow-based PDMS channel for all additional procedures: addition of competent cells, electroporation, addition of culture media, culture, fluorescence imaging and sorting.

Thermal regulation across the system was controlled by positioning peltier modules below the device to provide the 4°C, 37°C, 21°C and 50°C temperatures required for cell storage, cell culture and DNA assembly. Following droplet merger the gene parts were assembled at 21°C for Golden-Gate and 50°C for Gibson assembly, respectively. Gene delivery into electrocompetent cells was then accomplished by providing two 200 ms pulses of 1800 V/ cm by the same electrodes employed for digital microfluidics. Following addition of culture media and 24 hrs culture, successfully transformed cells exhibited a purple color and strong red autofluorescence, indicative of the production of violacein. The flexibility of digital microfluidics, PDMS valves and peltier modules affords quick optimization of operating protocols. The merger of these technologies provides us with a platform, which can perform a variety of operations with minimal electrical and fluidic inputs. This technology will be of great utility for systematic interpretation of gene delivery methods and high-throughput screening of gene variants with minimal reagent requirements.

1. S. C. Shih, G. Goyal, P. W. Kim, N. Koutsoubelis, J. D. Keasling, P. D. Adams, N. J. Hillson and A. K. Singh, *ACS Synth Biol*, 4(10) 2015, 1151-1164.
2. P. C. Gach, S. C. Shih, J. Sustarich, J. D. Keasling, N. J. Hillson, P. D. Adams and A. K. Singh, *ACS Synth Biol*, 5(5) 2016, 426-433.
3. S. Y. Choi, K. H. Yoon, J. I. Lee and R. J. Mitchell, *Biomed Res Int*, 2015, 2015, 465056.
4. C. J. Balibar and C. T. Walsh, *Biochemistry*, 2006, 45, 15444-15457.

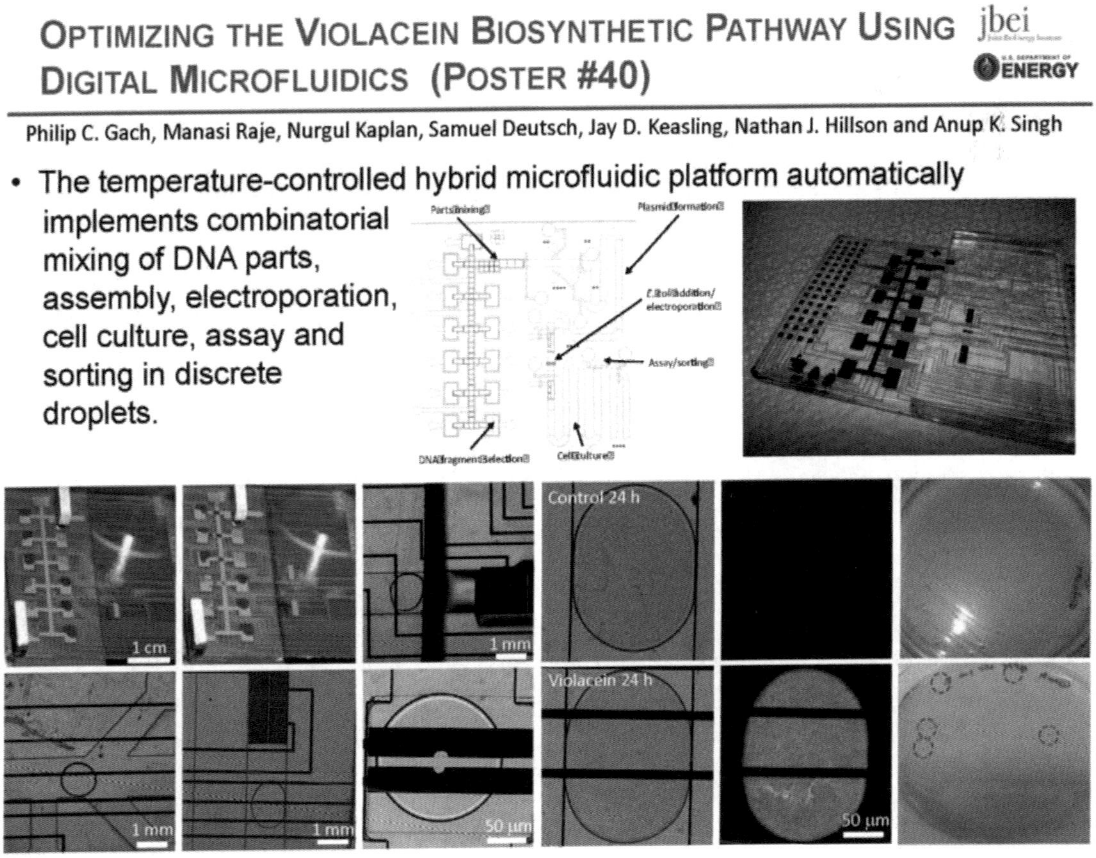

Overcoming the Membrane Barrier: Recruitment of ?-Glutamyl Transferase for Intracellular Release of Cargo Molecules

- Conference: Synthetic Biology Engineering Evolution Design SEED
- Year: 2017
- Proceeding: 2017 Synthetic Biology: Engineering, Evolution & Design (SEED)
- Group: Poster Session
- Session:

  Confirmed Posters

Authors:
Kuenzl, T., ETH Zurich
Panke, S., ETH Zurich
Marlière, P.
Herdewijn, P.
Srivastava, P.
Sroka, M., ETH Zurich

Small molecule-inducible transcription factors, or sensors, enable temporal control of toxic genes, provide the inputs forÂ *in vivo*Â computation, and allow a host to sense and respond to its environment. In this ongoing work we endeavor to improve the characteristics of several sensors using a mixture of rational engineering and directed evolution. We have improved the dynamic range, sensitivity, and/or specificity of 18 sensors. As an example of the utility of these sensors, we have used five optimized sensors to independently control the expression of five components of a metabolic pathway. By varying the amount of each inducer, we can easily and rapidly optimize the expression profile of the pathway without the need to construct large libraries. To facilitate the adoption of multiplex, sensor-based genetic control, we have created a stable cell-line capable of responding to 11 inputs, thus enabling the modular control of multiple genetic outputs with minimal optimization.

Paper-Based Detection with Synthetic Gene Circuits

- Conference: Synthetic Biology Engineering Evolution Design SEED
- Year: 2017
- Proceeding: 2017 Synthetic Biology: Engineering, Evolution & Design (SEED)
- Group: Poster Session
- Session:

  Confirmed Posters

Authors:
Lux, M.
Plants are a rich source of unique scaffolds, including 25% of natural-product-derived drugs. However, the discovery, synthesis, and overall material supply chains for sourcing plant natural products and their derivatives remain ad hoc, biased, and tedious. While microbial biosynthesis presents compelling alternatives to traditional approaches based on extraction from natural plant hosts, many challenges exist in the reconstruction of plant specialized metabolic pathways in microbial hosts. My laboratory has developed approaches to address the challenges that arise in the reconstruction of complex biosynthesis schemes, including spatial engineering strategies to direct the activities and specificities of pathway enzymes and recoding strategies to address folding, processing, and stability issues that may arise with the expression of plant enzymes in heterologous microbial hosts. We have utilized these strategies to develop yeast-based production platforms for an important class of plant alkaloids, the benzylisoquinoline alkaloids, including the medicinal opioids. These synthetic biology platforms will lead to transformative advances in natural product discovery, drug development, and production.

Pbnj Gibson Cloning for Fast Turn-Around Donor Constructs in Gene-Editing Applications

- Conference: Synthetic Biology Engineering Evolution Design SEED
- Year: 2017
- Proceeding: 2017 Synthetic Biology: Engineering, Evolution & Design (SEED)
- Group: Poster Session
- Session:

  Confirmed Posters

Authors:
Turner, I. K.
Chen, C.

Synthetic Biology aims to implement new cellular features through the design of original genetic parts and systems. A wide variety of circuit architectures have been built successfully, proving the ability of synthetic genetic parts to predictably interact with each other within a single bacterial cell [4]. The next stage of synthetic biology will involve the *de novo* engineering of self-organizing multicellular systems. One of the main methods found in nature for creating complex spatially distributed systems is cellular differentiation through asymmetric cell division (ACD), whereby stem cells spawn off irreversibly differentiated daughters. The presence of different cell lineages within an eukaryotic tissue allows for the partitioning of metabolic and structural tasks. Furthermore, if one or both of the daughter cells becomes incapable of dividing, controlling asymmetric division probability can lead to tissue size and shape control [5]. In complex eukaryotic organisms ACD is typically brought about by the differential expression of key genes within each cell type. Here, we engineered a novel method for generating ACD in *Escherichia coli* and regulating its frequency within the population. This work provides a platform for the construction of complex synthetic tissues and organisms from bacterial cells that differentiate during development of the population.

Current methods for creating differentiated cell types in synthetic bacteria utilize gene regulatory networks that allow for bistable gene expression [2, 1]. While such systems generate bimodal populations, they are very sensitive to both intrinsic and extrinsic noise sources that can randomly force transitions between states [3]. We use a novel method for creating terminally differentiated synthetic bacterial cells. Specifically, we refactored DNA shuttling mechanisms from *Caulobacter crescentus*

into *E. coli* to target plasmid DNA into one and only one daughter cell upon division. Thus, the two daughter cells will not just have different gene expression states, but will be irreversibly genetically distinct. Our findings will lead to synthetic "tissues" that can be used in complex settings.

References

[1] R. G. Egbert and E. Klavins. Fine-tuning gene networks using simple sequence repeats.*Proceedings of the National Academy of Sciences*, 109(42):16817â“16822, 2012.

[2] T. S. Gardner, C. R. Cantor, and J. J. Collins. Construction of a genetic toggle switch in *Escherichia coli*. *Nature*, 403(6767):339â“342, 2000.

[3] C. Gupta, J. M. LÃ³pez, W. Ott, K. JosiÄ‡, and M. R. Bennett. Transcriptional delay stabilizes bistable gene networks. *Physical review letters*, 111(5):058104, 2013.

[4] A. S. Khalil and J. J. Collins. Synthetic biology: applications come of age. *Nature Reviews*

Genetics, 11(5):367â“379, 2010.

[5] A. D. Lander, K. K. Gokoffski, F. Y. Wan, Q. Nie, and A. L. Calof. Cell lineages and the logic of proliferative control. *PLoS Biol*, 7(1):e1000015, 2009.

Phoenix: A Systems Engineering Approach to Spatial and Temporal Pattern Synthesis in Genetic Systems

- Conference: Synthetic Biology Engineering Evolution Design SEED
- Year: 2017
- Proceeding: 2017 Synthetic Biology: Engineering, Evolution & Design (SEED)
- Group: Poster Session
- Session:

  Confirmed Posters

Authors:
Madsen, C., Boston University
Appleton, E., Harvard Medical School
Vaidyanathan, P., Boston University
Pacheco, A.
Vasile, C. I.
Haghighi, I.
Mardian, R.
Roehner, N., University of Utah
Belta, C.
Densmore, D., Boston University

Engineered biological systems hold great potential to transform technologies in health, medicine and sustainability. Challenges in the design of biological devices include our incomplete understanding of the sequence-function relationships underlying the design space, and the limited throughput of experiments required to characterize individual devices. To significantly improve the efficiency of engineering highly functional gene-regulatory RNA devices, we 1) developed a massively parallel FACS-seq screen to assay the activity of hundreds of thousands of ribozyme switch library variants in yeast and mammalian cells, 2) adapted and automated a similarly high-throughput *in vitro* Cleave-seq assay on a liquid handler for parallelization and rapid iteration of the design-build-test cycle, and 3) employed computational analyses to extract design principles from the large datasets to inform future rational design.

To assay their activity in yeast or mammalian cells, ribozyme switches - a class of RNA devices with ligand-controllable self-cleavage activity - are integrated into the untranslated region of a fluorescent reporter, such that cleavage directly modulates fluorescence levels. Combining fluorescence-activated cell sorting (FACS) and next-generation sequencing (NGS), we extensively interrogated the effect of tertiary loop I-loop II interaction motifs on ribozyme activity, characterizing 10^5 variants in a single

FACS-seq experiment[1]. Cleave-seq measures *in vitro* cleavage activities of the ribozyme switches at a similar throughput using NGS, and automation of this method allowed us to quickly iterate through different library designs, while obtaining insight into differences in conserved sequence motifs between *in vitro* experiments and inside cells. With these high throughput techniques, we generated new RNA switches to a number of small molecule therapeutics and metabolites, which can serve as gene-control devices and live cell genetic sensors. Our analysis of the large datasets also identified a set of general sequence motifs that result in enhanced ribozyme cleavage activity, highlighting the importance of pseudoknot-like base pairs at the base of the ribozyme stem loops to mediate tertiary interactions. These datasets serve as a rich resource to further enable the discovery of novel hammerhead ribozymes with bioinformatics approaches and streamline the design of RNA switches for synthetic biology applications.

[1]Townshend, B., Kennedy, A. B., Xiang, J. S., & Smolke, C. D. (2015). High-throughput cellular RNA device engineering. Nature methods, 12(10), 989-994.

Probing Enzyme Sequence-Function Relationships By Deep Mutational Scanning of Two Enzyme Homologs

- Conference: Synthetic Biology Engineering Evolution Design SEED
- Year: 2017
- Proceeding: 2017 Synthetic Biology: Engineering, Evolution & Design (SEED)
- Group: Poster Session
- Session:

  Confirmed Posters

Authors:
Chen, J., UBC
Socha, R., University of British Columbia
Life, B. E., University of British Columbia
Tokuriki, N., University of British Columbia
Fowler, D., University of Washington

Synthetic biology experiments require optimization of pathways consisting of many genes and other genetic elements and given the large number of alternatives available for each element, optimization of a pathway can require large number of experiments. Currently, these experiments are done using fairly large amounts of costly reagents per experiment making the process very expensive, extremely slow and irreproducible. Our lab has previously developed droplet-based microfluidic systems for automating specific procedures of synthetic biology: gene assembly and electroporation [1] or heat-shock and culture [2]. However, these operations were performed separately and required additional manual sample preparation to perform the proof-of-principle experiments.

The current system integrates these processes to provide a programmable device for optimizing biosynthetic pathways. We applied this platform to screen for the production of violacein, a molecule with applications in textiles, agriculture and found to have antibacterial, antiviral and anticancer behaviors [3]. Violacein can be biosynthesized in *E. coli* from L-tryptophan in a process using five enzymes [4]. Digital microfluidics was initially used for dispensing and mixing the gene variants, vector backbone and assembly reagents because of the technologies great control of individual droplets and programmability. This component of the device is capable of generating a combinatorial library composed of 25 variants of the 11 kB violacein producing plasmid VioABCDE. Following combinatorial mixing of gene parts the droplets were transferred to a flow-based PDMS channel for all additional procedures: addition of competent cells, electroporation, addition of culture media, culture, fluorescence imaging and sorting.

Thermal regulation across the system was controlled by positioning peltier modules below the device to provide the 4°C, 37°C, 21°C and 50°C temperatures required for cell storage, cell culture and DNA assembly. Following droplet merger the gene parts were assembled at 21°C for Golden-Gate and 50°C for Gibson assembly, respectively. Gene delivery into electrocompetent cells was then accomplished by providing two 200 ms pulses of 1800 V/ cm by the same electrodes employed for digital microfluidics. Following addition of culture media and 24 hrs culture, successfully transformed cells exhibited a purple color and strong red autofluorescence, indicative of the production of violacein. The flexibility of digital microfluidics, PDMS valves and peltier modules affords quick optimization of operating protocols. The merger of these technologies provides us with a platform, which can perform a variety of operations with minimal electrical and fluidic inputs. This technology will be of great utility for systematic interpretation of gene delivery methods and high-throughput screening of gene variants with minimal reagent requirements.

1. S. C. Shih, G. Goyal, P. W. Kim, N. Koutsoubelis, J. D. Keasling, P. D. Adams, N. J. Hillson and A. K. Singh, *ACS Synth Biol*, 4(10) 2015, 1151-1164.
2. P. C. Gach, S. C. Shih, J. Sustarich, J. D. Keasling, N. J. Hillson, P. D. Adams and A. K. Singh, *ACS Synth Biol*, 5(5) 2016, 426-433.
3. S. Y. Choi, K. H. Yoon, J. I. Lee and R. J. Mitchell, *Biomed Res Int*, 2015, 2015, 465056.
4. C. J. Balibar and C. T. Walsh, *Biochemistry*, 2006, 45, 15444-15457.

Probiotic Design through Microbial Community Modelling Using Genome-Scale Metabolic Models

- Conference: Synthetic Biology Engineering Evolution Design SEED
- Year: 2017
- Proceeding: 2017 Synthetic Biology: Engineering, Evolution & Design (SEED)
- Group: Poster Session
- Session:

  Confirmed Posters

Authors:
Babaei, P., Chalmers University of Technology
We have engineered *Escherichia coli* to overproduce aliphatic methyl ketones (MK) in the C_{11} to C_{15} (diesel) range; this group of MK includes 2-undecanone and 2-tridecanone, which have favorable cetane numbers and are also of importance to the flavor and fragrance industry. Overall, we have made specific improvements that resulted in 50,000-fold enhancement in MK titer relative to that of a fatty acid-overproducing *E. coli* strain. The first generation of engineered *E. coli* (Goh et al. 2012) had modifications that included: (a) overproduction of Î²-ketoacyl-coenzyme A (CoA) thioesters achieved by modification of the Î²-oxidation pathway (specifically, overexpression of a heterologous acyl-CoA oxidase and native FadB, and chromosomal deletion of *fadA*) and (b) overexpression of a native thioesterase (FadM). Subsequent genetic modifications included balancing overexpression of *fadR* and *fadD* to increase fatty acid flux into the pathway, consolidation of the pathway from two plasmids into one, codon optimization, and knocking out key acetate production pathways (resulting in strain EGS1895). This strain produced an MK titer of 1.4 g/L with 1% glucose in shake-flask experiments, representing 40% of the maximum theoretical yield. ^{13}C-based metabolic flux analysis with EGS1895 showed increased carbon flux through the pentose phosphate pathway (PPP), which we hypothesized is to generate more NADPH - an important co-factor for FabG (Î²-ketoacyl-ACP reductase) activity during fatty acid biosynthesis.Â In an attempt to divert flux away from PPP, we overexpressed alternate FabG variants that we have demonstrated preferentially use NADH instead of NADPH. Specifically, overexpression of the NADH-dependent *Acholeplasma laidlawii* FabG increased MK production by 20-50% in shake-flask experiments and had more sustained MK production in fed-batch fermentation compared to EGS1895. We have achieved MK titers of ~6.0 g/L in fed-batch fermentation, which is the best MK titer observed to date.

Production of Seleno-scFvs with Defined Structure in E. coli

- Conference: Synthetic Biology Engineering Evolution Design SEED
- Year: 2017
- Proceeding: 2017 Synthetic Biology: Engineering, Evolution & Design (SEED)
- Group: Poster Session
- Session:

  Confirmed Posters

Authors:
Thyer, R., University of Texas at Austin

Microbes can be engineered to produce biologically-derived replacements for gasoline, diesel, and jet fuel. However, a major technical challenge is that cells can only tolerate limited concentrations of typical fuels. Hydrocarbons are toxic because they damage the cell membrane and increase membrane fluidity. Thus, without further engineering there is a fundamental upper limit to how much biofuel microbes can produce, and cells must trade off production against survival. However, there are many natural examples of microbes that thrive in oil rich environments, such as around natural oil seeps or near oil spills. These microbes harbor mechanisms for tolerating the toxic effects of hydrocarbons. By using library-based approaches, we have identified several novel strategies for improving microbial biofuel tolerance by selecting for genes that enable survival. For example, expression of efflux pumps can dramatically improve tolerance in production strains, and we have further improved survival by designing synthetic feedback systems to control pump expression. This talk will discuss our efforts to enhance microbial synthesis of next-generation biofuels by developing tools for improving microbial tolerance of biofuel production conditions.

Protein Engineering a Carbon-Neutral Photorespiration Pathway

- Conference: Synthetic Biology Engineering Evolution Design SEED
- Year: 2017
- Proceeding: 2017 Synthetic Biology: Engineering, Evolution & Design (SEED)
- Group: Poster Session
- Session:

  Confirmed Posters

Authors:
Trudeau, D.
Tawfik, D., The Weizmann Institute of Science, Rehovot, Israel
Our ability to predict protein expression from DNA sequence alone remains poor, reflecting our limited understanding of cis-regulatory grammar and hampering the design of engineered genes for synthetic biology applications. We have generated a model that predicts the translational efficiency of the 5â□™ untranslated region (UTR) of mRNAs in the yeast *Saccharomyces cerevisiae.* We constructed a library of nearly half a million 50 nucleotide-long random 5â□™ UTRs and assayed them in a single massively parallel growth selection experiment. The resulting data have allowed us to quantify the impact on translation of Kozak sequence composition, upstream open reading frames (uORFs) and secondary structure. With this data, we have trained a convolutional neural network on the random library and validated it by predicting the translational efficiency of 5â□™ UTRs that natively occur in yeast. The model additionally was used to computationally evolve highly translating 5â□™ UTRs. We have confirmed experimentally that the great majority of the evolved sequences lead to higher translation rates than the starting sequences, demonstrating the predictive utility of this model.

Protein Engineering and Precise Modulation of Protein Expression in the Protist Parasite Trypanosoma Brucei

- Conference: Synthetic Biology Engineering Evolution Design SEED
- Year: 2017
- Proceeding: 2017 Synthetic Biology: Engineering, Evolution & Design (SEED)
- Group: Poster Session
- Session:

  Confirmed Posters

Authors:
Pinger, J., Rockefeller University
Papavasiliou, F. N.

Our ability to predict protein expression from DNA sequence alone remains poor, reflecting our limited understanding of cis-regulatory grammar and hampering the design of engineered genes for synthetic biology applications. We have generated a model that predicts the translational efficiency of the 5â□™ untranslated region (UTR) of mRNAs in the yeast *Saccharomyces cerevisiae.* We constructed a library of nearly half a million 50 nucleotide-long random 5â□™ UTRs and assayed them in a single massively parallel growth selection experiment. The resulting data have allowed us to quantify the impact on translation of Kozak sequence composition, upstream open reading frames (uORFs) and secondary structure. With this data, we have trained a convolutional neural network on the random library and validated it by predicting the translational efficiency of 5â□™ UTRs that natively occur in yeast. The model additionally was used to computationally evolve highly translating 5â□™ UTRs. We have confirmed experimentally that the great majority of the evolved sequences lead to higher translation rates than the starting sequences, demonstrating the predictive utility of this model.

Proteomics-Based Evaluation of Cell Free Protein Synthesis

- Conference: Synthetic Biology Engineering Evolution Design SEED
- Year: 2017
- Proceeding: 2017 Synthetic Biology: Engineering, Evolution & Design (SEED)
- Group: Poster Session
- Session:

  Confirmed Posters

Authors:
Doktycz, M. J., Oak Ridge National Laboratory
Mohr, B. P., Oak Ridge National Laboratory
Garcia, D. C., University of Tennessee, Knoxville
Morrell-Falvey, J. L., Oak Ridge National Laboratory
Standaert, R. F.
Hurst, G. B., Oak Ridge National Laboratory

DNA assembly forms the cornerstone of modern synthetic biology. Despite the numerous available methods, scarless multi-fragment assembly of large plasmids remains challenging. Furthermore, the upcoming wave in molecular biological automation demands a rethinking of how we perform DNA assembly. To streamline automation workflow and minimize operator intervention, a non-enzymatic assembly method is highly desirable. Here, we report the optimization and operationalization of a process called Twin-Primer Assembly (TPA), which is a method to assemble PCR-amplified fragments into a plasmid without the use of enzymes.

Each fragment used in TPA requires two PCR products, amplified separately using two sets of primers. One product has an overlap to the fragment that comes before and the other has an overlap to the fragment that comes after. The length of the overlap region is determined by the melting temperature (T_M) of the overlap region, and all overlaps used in an assembly are designed to have the same T_M. The two PCR products are mixed and fully denatured at high temperature before they are allowed to re-anneal slowly, generating intermediates that have overhangs on both sides. These intermediates are then allowed to hybridize at an elevated temperature to form a circular nicked plasmid. Upon *E. coli* transformation, the cells phosphorylate and ligate the nicks to create the final product plasmid.

The TPA non-enzymatic DNA assembly method is scarless and generally sequence independent. It is capable of assembling a 7 kb plasmid from 10 fragments at close to 80% fidelity, and a 31kb plasmid from 5 fragments at around 50% fidelity. Assembling more than 10 fragments is likely possible for constructing smaller plasmids, while assembling a plasmid larger than 31 kb is likely possible using fewer fragments. Even without the use of enzymes, the performance of TPA is on par with some of the best *in vitro* assembly methods currently available. TPA should be an invaluable addition to a synthetic biologistâ□™s toolbox.

Putative DNA-Binding Proteins Affecting the Subcellular Organization of Bacillus Subtilis

- Conference: Synthetic Biology Engineering Evolution Design SEED
- Year: 2017
- Proceeding: 2017 Synthetic Biology: Engineering, Evolution & Design (SEED)
- Group: Poster Session
- Session:

  Confirmed Posters

Authors:
Hartman, S. E., Texas A&M University

Bacteria lack compartmentalization and therefore rely on subcellular organization to temporally and spatially localize major cellular processes. As one of the major structures in the cell, the nucleoid possesses positional cues that help direct important functions such as DNA replication and cell division. While some of the factors that contribute to maintaining subcellular organization are characterized, many of the underlying determinants are still unknown. In *Bacillus subtilis*, there remains a class of putative DNA-binding proteins currently uncharacterized, and I hypothesize that some of these proteins use DNA to localize functions important in maintaining subcellular organization. Since many uncharacterized genes lack easily discernable knockout phenotypes, I utilized a gene misexpression approach to identify DNA-binding proteins capable of perturbing some aspect of cellular organization. The misexpression library employed comprises over 800 strains, each containing an uncharacterized gene under the control of an inducible promoter on the *B. subtilis* chromosome. Genes for 98 putative DNA-binding proteins were misexpressed and analyzed for their ability to grow in the presence of inducer. Of these strains, nine exhibited reduced growth and 17 strains did not grow. When analyzed by fluorescence microscopy, 19 of these strains could be associated with a distinct phenotype, including defects in the structure and placement of the chromosome, abnormal cell division, and loss of cell shape. To further characterize strains with the most severe phenotypes we performed RNA-seq experiments to determine if gene misexpression alters the transcription profile. We identified four genes that do not affect transcription, suggesting that the misexpression phenotype is not due to transcriptional regulation. We hypothesize that this class of DNA-binding proteins act as positional cues that help maintain subcellular organization.

Putative DNA-Binding Proteins Affecting the Subcellular Organization of *Bacillus subtilis* #49

Sarah Hartman and Jennifer Herman, Texas A&M University

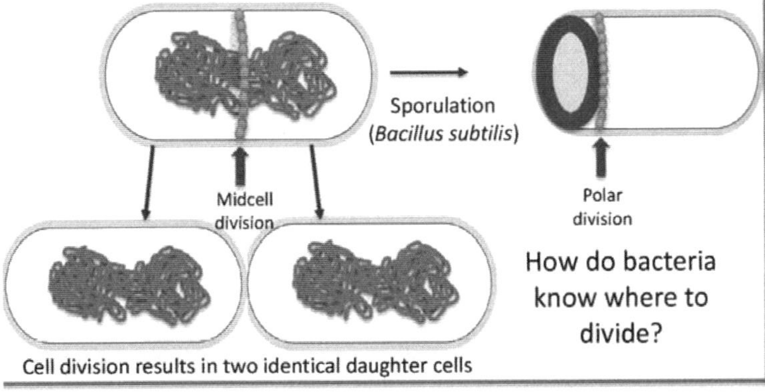

How do bacteria know where to divide?

How do bacteria regulate processes in time and space?

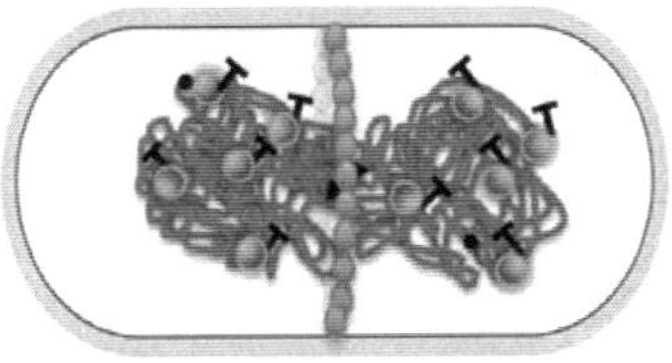

DNA localizes functions throughout the cell

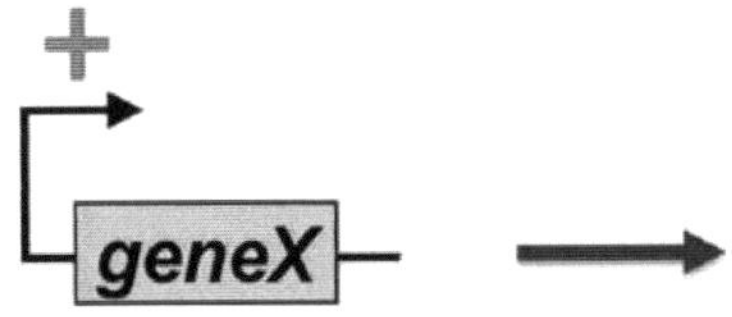

Screen a misexpression library for gene products that affect subcellular organization by binding DNA

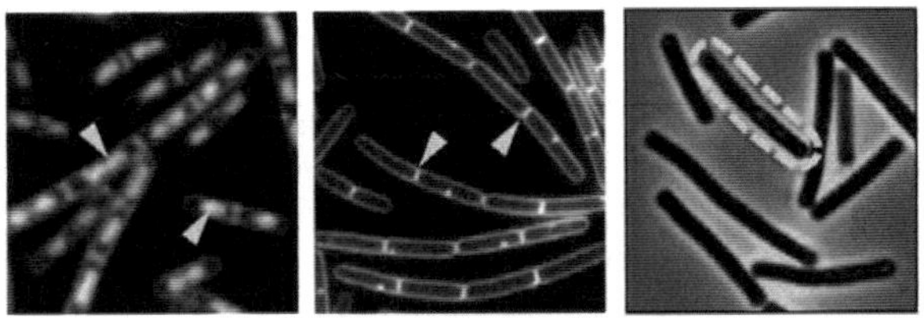

DNA Replication | Cell Division | Cell Shape

Quantifying Noise Correlations in the Multiple Antibiotic Resistance (Mar) Network

- Conference: Synthetic Biology Engineering Evolution Design SEED
- Year: 2017
- Proceeding: 2017 Synthetic Biology: Engineering, Evolution & Design (SEED)
- Group: Poster Session
- Session:

  Confirmed Posters

Authors:
El Meouche, I., University of Vermont
Rossi, N. A., Boston Univeristy
Dunlop, M. J., N/A

Heterologously expressingÂ *Shewanella oneidensis*Â MR-1Â cytochromes *c* (cyt *c*) inÂ *Escherichia coli* enables cellular states to be monitored and controlled through electron flow. However, the electron flux in *E. coli* is significantly less than *Shewanella oneidensis* MR-1, thus limiting the extent of our electronic control. We hypothesize that differences in the cytochrome *c* maturation (*ccm*) machinery between these two species may be responsible for the lower expression levels of cyt *c* and extracellular electron transfer (EET) efficiency. In this research, we constructed random mutations within *ccm* genes by prone error PCR and screened for increases in cyt *c* production in *E. coli* in 96-well plates. Two ccmH mutation strains were shown to exhibit increased cyt *c* expression. Since these two mutations occurred in the C-terminal domain of *ccmH*, which has a similar structure of *ccmI* in the *Shewanella oneidensis* MR-1, we also constructed a hybrid *ccmH* comprising the N-terminal domain of *Escherichia coli* and the C-terminal domain of *Shewanella oneidensis* MR-1 ($ccmH^N$:ccmI). Indeed, cells from this engineered $ccmH^N$:ccmI strain possess a redder color and higher expression of cyt *c*. We measured the electron flux out of these modified variants using three-electrode microbial electrochemical experiments and showed that the $ccmH^N$:ccmI strain produces ~40% more current per cell. Thus, modulating maturation of cyt *c* is not only possible, but can improve electron transfer in synthetic biological systems.

Quantitative Modeling of Integrase Dynamics Using a Novel Python Toolbox for Parameter Inference

- Conference: Synthetic Biology Engineering Evolution Design SEED
- Year: 2017
- Proceeding: 2017 Synthetic Biology: Engineering, Evolution & Design (SEED)
- Group: Poster Session
- Session:

  Confirmed Posters

Authors:
Swaminathan, A., California Institute of Technology
Hsiao, V., Amyris
Murray, R. M., California Institute of Technology

Cell surface glycans make up an important mode of intercellular communication and the enzymes that assemble and degrade them are not only important targets for therapeutics, but also powerful tools for chemical glycobiology. Our recent work has explored 1) the use of engineered glycoside hydrolases for generating universal blood by antigen removal, 2) the identification of inhibitors to glycosyltransferases in cancer pathways, and 3) the development of engineered â□œglycosynthasesâ□□ to assemble and probe the function of glycosaminoglycan structures.

We have previously demonstrated that blood antigen-cleaving glycoside hydrolases can be engineered as tools towards the generation of universal blood through the cleavage of carbohydrate blood group antigens. Identification and directed evolution of these enzymes is enabled by a strategy that involves high-throughput screening with a sensitive fluorescence based assay. Using a similar approach, creatively repurposed to assay for glycosyltransferase rather than glycoside hydrolase activity, we have developed a high-throughput screen for inhibitors targeting fucosyltransferase enzymes that are highly expressed in cancers. Such inhibitors have potential as therapeutics for certain cancers wherein these enzymesâ□™ upregulated activity results in the assembly of carbohydrate structures (e.g. sialyl Lewis^{X}) involved in malignant processes including metastasis. In another focus, we are developing tools for chemoenzymatic glycan synthesis through mechanism-based engineering of â□œglycosynthaseâ□□ enzyme biocatalysts to assemble specific glycosaminoglycan structures. A facile enzymatic synthesis of these glycans will enable glycobiologists to probe their interactions in processes such as neural development and cancer.

Rapid Identification and Quantitative Assessment of Plant Enhancers in Full-Genome Libraries

- Conference: Synthetic Biology Engineering Evolution Design SEED
- Year: 2017
- Proceeding: 2017 Synthetic Biology: Engineering, Evolution & Design (SEED)
- Group: Poster Session
- Session:

  Confirmed Posters

Authors:
Dorrity, M. W., University of Washington
Our ability to predict protein expression from DNA sequence alone remains poor, reflecting our limited understanding of cis-regulatory grammar and hampering the design of engineered genes for synthetic biology applications. We have generated a model that predicts the translational efficiency of the 5â□™ untranslated region (UTR) of mRNAs in the yeast *Saccharomyces cerevisiae*. We constructed a library of nearly half a million 50 nucleotide-long random 5â□™ UTRs and assayed them in a single massively parallel growth selection experiment. The resulting data have allowed us to quantify the impact on translation of Kozak sequence composition, upstream open reading frames (uORFs) and secondary structure. With this data, we have trained a convolutional neural network on the random library and validated it by predicting the translational efficiency of 5â□™ UTRs that natively occur in yeast. The model additionally was used to computationally evolve highly translating 5â□™ UTRs. We have confirmed experimentally that the great majority of the evolved sequences lead to higher translation rates than the starting sequences, demonstrating the predictive utility of this model.

Rational Design of Ribozyme Switches through Big Data and Machine Learning

- Conference: Synthetic Biology Engineering Evolution Design SEED
- Year: 2017
- Proceeding: 2017 Synthetic Biology: Engineering, Evolution & Design (SEED)
- Group: Poster Session
- Session:

  Confirmed Posters

Authors:
Schmidt, C. M., Stanford University
Smolke, C. D., Stanford University

RNA switches are a class of functional RNAs that support conditional regulation of gene expression. These genetic switches have broad application in the design of biological systems responsive to changes in their environment with implications for the fields of metabolic engineering, environmental sensing, and therapeutics. The mechanisms underlying the activity of RNA switches rely on programmed structural conformational changes encoded in the sequence of the RNA molecule, implying that an improved elucidation of relationships between sequence, structure, and function can be used to improve our capacities for de novo design. One type of RNA switch is the ribozyme switch, in which binding of the target ligand changes the structure and thus cleavage activity of the switch. Here, we leverage sequencing data and machine learning methods to relate the sequence and structure of RNA molecules to their function and develop models that predict the function of novel ribozyme switches. Through prior work, we have access to data on the activity of hundreds of thousands of ribozyme switch sequences. Using automated structural analysis and machine learning methods, we leveraged a wealth of sequence-activity data on hundreds of thousands of ribozyme switches to develop mathematical models that predict the in vivo activity of a given ribozyme based on its sequence. We have used these models for the de novo design of novel ribozyme switches that exhibit changes in in vivo gene-regulatory activity upon introduction of a target ligand. Our work demonstrates that computational tools can be used to convert RNA sequences into quantitative features that represent key sequence and structural motifs and that these features can be used to develop models that can learn how interactions between these motifs translates to the activity of the functional RNA molecule. The predictive models we developed provide a broadly accessible tool for designing custom biosensors for use in engineering novel biological systems

Rational design of tertiary ribozyme switches through big data and machine learning

Calvin Schmidt
Stanford University
Poster 128

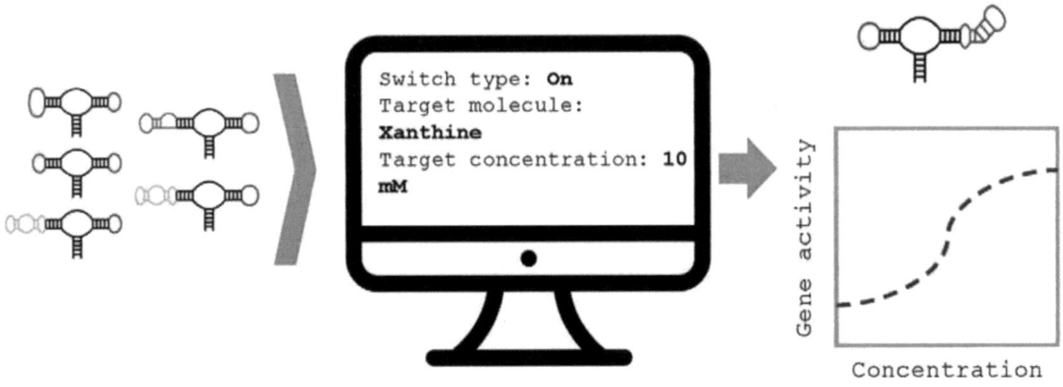

Rational Mutagenesis to Enable Yeast DNA Polymerase ? (POL3) Tobecome a Calcium-Sensitive Recording Device

- Conference: Synthetic Biology Engineering Evolution Design SEED
- Year: 2017
- Proceeding: 2017 Synthetic Biology: Engineering, Evolution & Design (SEED)
- Group: Poster Session
- Session:

  Confirmed Posters

Authors:
Biggs, B. W., University of Southern California
Tyo, K. E. J., Northwestern University
Bhan, N., Rensselear Polytechnic Institute
de Paz, A., Northwestern University
Cybulski, T., Northwestern University

Improving the RNA switch design process through big data and machine learning

Calvin M. Schmidt, Christina D. Smolke
Stanford University; Stanford, California

RNA switches are a class of functional RNAs that support conditional regulation of gene expression. These genetic switches have broad application in the design of biological systems responsive to changes in their environment with implications for the fields of metabolic engineering, environmental sensing, and therapeutics. The mechanisms underlying the activity of RNA switches rely on programmed structural conformational changes encoded in the sequence of the RNA molecule, implying that an improved elucidation of relationships between sequence, structure, and function can be used to improve our capacities for de novo design. One type of RNA switch is the ribozyme switch, in which binding of the target ligand changes the structure and thus cleavage activity of the switch. Here, we leverage sequencing data and machine learning methods to relate the sequence and structure of RNA molecules to their function and develop models that predict the function of novel ribozyme switches. Through prior work, we have access to data on the activity of hundreds of thousands of ribozyme switch sequences. Using automated structural analysis and machine learning methods, we leveraged a wealth of sequence-activity data on hundreds of thousands of ribozyme switches to develop mathematical models that predict the in vivo activity of a given ribozyme based on its sequence. We have used these models for the de novo design of novel ribozyme switches that exhibit changes in in vivo gene-regulatory activity upon introduction of a target ligand. Our work demonstrates that computational tools can be used to convert RNA sequences into quantitative features that represent key sequence and structural motifs and that these features can be used to develop models that can learn how interactions between these motifs translates to the activity of the functional RNA molecule. The predictive models we developed provide a broadly accessible tool for designing custom biosensors for use in engineering novel biological systems

Recent Breakthroughs in Enabling Technologies Platforms for Wheat

- Conference: Synthetic Biology Engineering Evolution Design SEED
- Year: 2017
- Proceeding: 2017 Synthetic Biology: Engineering, Evolution & Design (SEED)
- Group: Poster Session
- Session:

  Confirmed Posters

Authors:
Kathiria, P.
Diedhiou, J.
Burton-Hughes, K.
Randhawa, H.
Marillia, E.
Microbes can be engineered to produce biologically-derived replacements for gasoline, diesel, and jet fuel. However, a major technical challenge is that cells can only tolerate limited concentrations of typical fuels. Hydrocarbons are toxic because they damage the cell membrane and increase membrane fluidity. Thus, without further engineering there is a fundamental upper limit to how much biofuel microbes can produce, and cells must trade off production against survival. However, there are many natural examples of microbes that thrive in oil rich environments, such as around natural oil seeps or near oil spills. These microbes harbor mechanisms for tolerating the toxic effects of hydrocarbons. By using library-based approaches, we have identified several novel strategies for improving microbial biofuel tolerance by selecting for genes that enable survival. For example, expression of efflux pumps can dramatically improve tolerance in production strains, and we have further improved survival by designing synthetic feedback systems to control pump expression. This talk will discuss our efforts to enhance microbial synthesis of next-generation biofuels by developing tools for improving microbial tolerance of biofuel production conditions.

Relative Fluorescent Units Considered Harmful

- Conference: Synthetic Biology Engineering Evolution Design SEED
- Year: 2017
- Proceeding: 2017 Synthetic Biology: Engineering, Evolution & Design (SEED)
- Group: Poster Session
- Session:

  Confirmed Posters

Authors:
Haddock-Angelli, T., iGEM Foundation
Baldwin, G. S., Imperial College London
Beal, J., Raytheon BBN Technologies
Gershater, M., Synthace
Storch, M., Imperial College London
Dwijayanti, A., Imperial College
de Mora, K., iGEM Foundation
Lizarazo, M., iGEM Foundation
Rettberg, R., iGEM Foundation
Consortium, I. I. S., iGEM Competition
Plants are a rich source of unique scaffolds, including 25% of natural-product-derived drugs. However, the discovery, synthesis, and overall material supply chains for sourcing plant natural products and their derivatives remain ad hoc, biased, and tedious. While microbial biosynthesis presents compelling alternatives to traditional approaches based on extraction from natural plant hosts, many challenges exist in the reconstruction of plant specialized metabolic pathways in microbial hosts. My laboratory has developed approaches to address the challenges that arise in the reconstruction of complex biosynthesis schemes, including spatial engineering strategies to direct the activities and specificities of pathway enzymes and recoding strategies to address folding, processing, and stability issues that may arise with the expression of plant enzymes in heterologous microbial hosts. We have utilized these strategies to develop yeast-based production platforms for an important class of plant alkaloids, the benzylisoquinoline alkaloids, including the medicinal opioids. These synthetic biology platforms will lead to transformative advances in natural product discovery, drug development, and production.

Repressing Integrase Attachment Site Operation with CRISPR-Cas9 in E. coli

- Conference: Synthetic Biology Engineering Evolution Design SEED
- Year: 2017
- Proceeding: 2017 Synthetic Biology: Engineering, Evolution & Design (SEED)
- Group: Poster Session
- Session:

  Confirmed Posters

Authors:
Shur, A.
Murray, R. M., California Institute of Technology

Cell surface glycans make up an important mode of intercellular communication and the enzymes that assemble and degrade them are not only important targets for therapeutics, but also powerful tools for chemical glycobiology. Our recent work has explored 1) the use of engineered glycoside hydrolases for generating universal blood by antigen removal, 2) the identification of inhibitors to glycosyltransferases in cancer pathways, and 3) the development of engineered â□œglycosynthasesâ□□ to assemble and probe the function of glycosaminoglycan structures.

We have previously demonstrated that blood antigen-cleaving glycoside hydrolases can be engineered as tools towards the generation of universal blood through the cleavage of carbohydrate blood group antigens. Identification and directed evolution of these enzymes is enabled by a strategy that involves high-throughput screening with a sensitive fluorescence based assay. Using a similar approach, creatively repurposed to assay for glycosyltransferase rather than glycoside hydrolase activity, we have developed a high-throughput screen for inhibitors targeting fucosyltransferase enzymes that are highly expressed in cancers. Such inhibitors have potential as therapeutics for certain cancers wherein these enzymesâ□™ upregulated activity results in the assembly of carbohydrate structures (e.g. sialyl Lewis^{X}) involved in malignant processes including metastasis. In another focus, we are developing tools for chemoenzymatic glycan synthesis through mechanism-based engineering of â□œglycosynthaseâ□□ enzyme biocatalysts to assemble specific glycosaminoglycan structures. A facile enzymatic synthesis of these glycans will enable glycobiologists to probe their interactions in processes such as neural development and cancer.

Reprogramming Yeast Mating to Characterize Protein-Protein Interactions in High-Throughput

- Conference: Synthetic Biology Engineering Evolution Design SEED
- Year: 2017
- Proceeding: 2017 Synthetic Biology: Engineering, Evolution & Design (SEED)
- Group: Poster Session
- Session:

  Confirmed Posters

Authors:
Younger, D., University of Washington
Berger, S.
Baker, D., University of Washington
Klavins, E., University of Washington
3-hydroxypropanoic acid (3-HP) is an important biomass-derivable platform chemical that can be converted into a number of industrially relevant compounds. There have been several attempts to produce 3-HP from renewable sources in cell factories, focusing mainly on *Escherichia coli*, *Klebsiella pneumoniae* and *Saccharomyces cerevisiae*. Despite the significant progress made in this field, commercially exploitable large-scale production of 3-HP in microbial strains has still not been achieved. In this study we investigated the potential of *Bacillus subtilis* as a microbial platform for bioconversion of glycerol into 3-HP. Our recombinant *B. subtilis* strains overexpress the two-step heterologous pathway containing glycerol dehydratase and aldehyde dehydrogenase from *K. pneumoniae*. Genetic engineering, driven by *in silico* optimization, and optimization of cultivation conditions resulted in a 3-HP titer of 10 g/L, in a standard batch cultivation. Our findings provide the first report of successful introduction of the biosynthetic pathway for conversion of glycerol into 3-HP in *B. subtilis*. With this relatively high titer in batch, and the robustness of *B. subtilis* in high density fermentation conditions, we expect that our production strains may constitute a solid basis for commercial production of 3-HP.

RNA Regulators of Gene Expression for High-Throughput Strain Engineering

- Conference: Synthetic Biology Engineering Evolution Design SEED
- Year: 2017
- Proceeding: 2017 Synthetic Biology: Engineering, Evolution & Design (SEED)
- Group: Poster Session
- Session:

  Confirmed Posters

Authors:
Glasscock, C. J., Cornell University
Lucks, J. B., Cornell University

Plants are a rich source of unique scaffolds, including 25% of natural-product-derived drugs. However, the discovery, synthesis, and overall material supply chains for sourcing plant natural products and their derivatives remain ad hoc, biased, and tedious. While microbial biosynthesis presents compelling alternatives to traditional approaches based on extraction from natural plant hosts, many challenges exist in the reconstruction of plant specialized metabolic pathways in microbial hosts. My laboratory has developed approaches to address the challenges that arise in the reconstruction of complex biosynthesis schemes, including spatial engineering strategies to direct the activities and specificities of pathway enzymes and recoding strategies to address folding, processing, and stability issues that may arise with the expression of plant enzymes in heterologous microbial hosts. We have utilized these strategies to develop yeast-based production platforms for an important class of plant alkaloids, the benzylisoquinoline alkaloids, including the medicinal opioids. These synthetic biology platforms will lead to transformative advances in natural product discovery, drug development, and production.

RNA-Aptamer Sensors for Phenotype Screening of Engineered Microbial Pathways

- Conference: Synthetic Biology Engineering Evolution Design SEED
- Year: 2017
- Proceeding: 2017 Synthetic Biology: Engineering, Evolution & Design (SEED)
- Group: Poster Session
- Session:

  Confirmed Posters

Authors:
Burke, C., University of Washington
Carothers, J., University of Washington

The recent years have provided tremendous improvements in microbial pathway engineering through the design of complex control circuits, however industrial relevant titers are still difficult to obtain. One major bottleneck in pathway engineering is screening for small molecule production since only a small fraction of engineered pathway products (chromophores or fluorophores) can be screened using traditional high throughput methods. Here, we present an RNA-aptamer based sensor system to detect secreted metabolites from engineered pathways in a high throughput and cost efficient manner. Taking inspiration from the many examples of naturally occurring kinetically controlled RNA-based sensors, such as riboswitches, we have engineered a set of RNA-aptamer based biosensors that sense and respond to *p*-amino-phenylalanine (*p*-AF) co-transcriptionally. To our knowledge, this is the first time that an RNA-aptamer based sensor has been engineered deliberately to function under kinetic control. These sensors were screened *in silico* from approximately 1.5E5 unique sequences that each contain the same parental RNA aptamer sequence. We have developed a set of *p*-AF sensors that exhibit EC_{50} values between 1 â“ 0.5 mM, ideal for screening the production of *p*-AF producing *E. coli* strains with titers as low as 100 mg/L. Currently, we can detect *p*-AF in cell culture supernatant, and we are working towards optimizing a high throughput assay to screen *p*-AF production strain variants. To validate our computational pipeline and demonstrate the utility of this method, we are also generating sensors for tyrosine, another small molecule product that has been produced using the core of the engineered *p*-AF pathway.

RNA-Aptamer Sensors for Phenotype Screening of Engineered Microbial Pathways

#37

Cassandra R. Burke, James M. Carothers
University of Washington, Seattle WA 98115

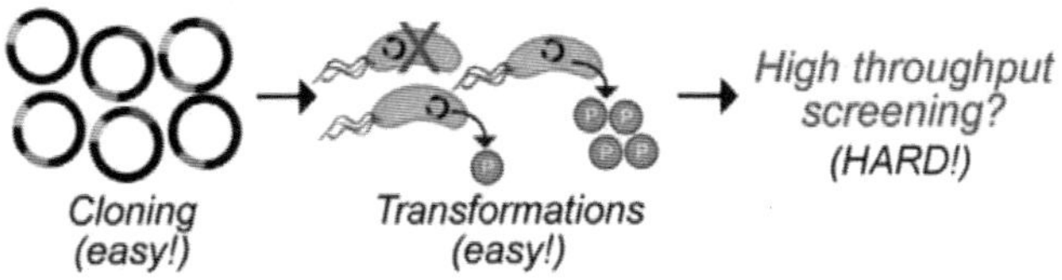

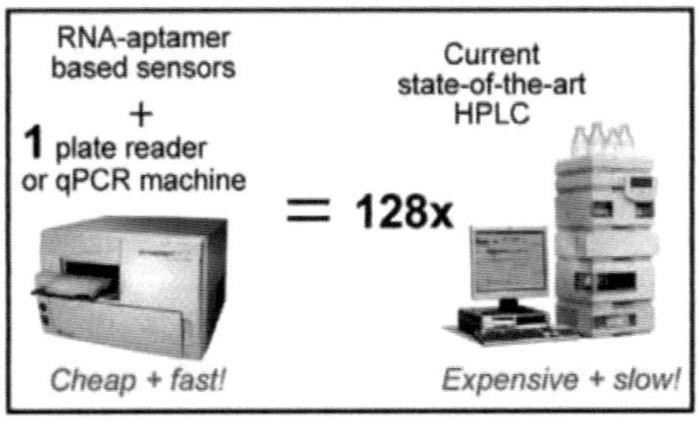

Ribosensor highlights:

Measure secreted small molecules
- Function in microbial cell supernatant
- Not susceptible to degradation
- Fast response time

Tunable set-points
- No re-engineering the aptamer
- Computationally tractable

Fast and inexpensive
- Computational pipeline
- No purification
- No HPLC

Points towards kinetic control

RNA-Guided Microbial Genome Editing Using Group II Introns and CRISPR-Cas Systems

- Conference: Synthetic Biology Engineering Evolution Design SEED
- Year: 2017
- Proceeding: 2017 Synthetic Biology: Engineering, Evolution & Design (SEED)
- Group: Poster Session
- Session:

  Confirmed Posters

Authors:
Zhang, Z.
McCann, C. H.
Davis, G. D.
Eastlund, E. R.
Amyris is an end-to-end biological manufacturer that uses synthetic biology to engineer microbes as catalysts for conversion of sugar into chemical products. Many of these chemical targets were previously derived from limited natural resources and economically inaccessible via traditional chemistries. In the last decade, Amyris has developed a sophisticated technology platform for engineering microbes to produce high levels of isoprenoids including artemisinin and farnesene. In recent years, we have begun to harness the power of this proven infrastructure to expand the portfolio of bio-manufactured chemical products.

In this talk, I will describe the current state-of-the-art tools at Amyris, including our automated strain engineering, high-throughput screening, and fermentation. Then I will discuss how these tools are evolving under the DARPA-funded â□œMgs to Kgsâ□□ project to enable the concurrent development of 100â□™s of products simultaneously. In the first year, the project has produced more than 150 target molecules from all parts of metabolic space. As the project continues, â□œMgs to Kgsâ□□ will enable Amyris to incorporate next generation high-throughput â□˜omics, automation, data science, and machine learning tools to further increase the rate at which biology can be engineered.

Saturation Mutagenesis of E. coli Sigma Factor As a Tool for Strain Engineering

- Conference: Synthetic Biology Engineering Evolution Design SEED
- Year: 2017
- Proceeding: 2017 Synthetic Biology: Engineering, Evolution & Design (SEED)
- Group: Poster Session
- Session:

  Confirmed Posters

Authors:
Park, J., Columbia University
Wang, H. H., Columbia University

Recent studies into the interplay between the microbiome and host have extensively implicated short-chain fatty acid (SCFA) fermentation products as contributors to both intra and extra intestinal health. In the context of the gut, butyrate (a four carbon SCFA) serves as the primary energy source for epithelial colonocytes, drives regulatory T cell differentiation and is therefore essential for the maintenance of colonic mucosal health. On a systemic level, butyrate may influence insulin sensitivity, cholesterol synthesis and autoimmunity. Typically, butyrate is produced by the action of slow-growing Gram positive organisms â□“ often Clostridial in origin â□“ during the fermentation of cellulose and complex carbohydrates. To counteract the depletion of butyrate in the gut, we iteratively engineered *Escherichia coli* Nissle 1917 to produce butyrate. The new butyrate-producing Nissle was capable of millimolar level production of butyrate comparable with the levels produced by enteric Clostridial species. The circuitry was placed under anaerobic control to support induction upon entry into the gastrointestinal track. Finally, proof-of-concept *in vivo* experimentation showed production of butyrate in the mouse gut. This engineered strain may be useful for the treatment of a variety of gastrointestinal and systemic conditions associated with SCFA deficiency.

Scaling up DNA Data Storage

- Conference: Synthetic Biology Engineering Evolution Design SEED
- Year: 2017
- Proceeding: 2017 Synthetic Biology: Engineering, Evolution & Design (SEED)
- Group: Poster Session
- Session:

  Confirmed Posters

Authors:
Takahashi, C., University of Washington
Dumas Ang, S.
Organick, L.
Chen, Y. J.
Lopez, R.
Yekhanin, S., Mircosoft Research
Makarychev, K.
Racz, M. Z.
Newman, S.
Nguyen, B.
Parker, H. Y.
Rashtchian, C.
Stewart, K.
Gupta, G.
Carlson, R.
Mulligan, J.
Carmean, D., Dcarmean@Microsoft.Com
Seelig, G., University of Washington
Ceze, L.
Strauss, K.
Protein engineering has important applications that range from developing sensors in bacteria to producing antibodies for cancer therapeutics. Re-engineering protein function usually requires many amino acid changes, which results in libraries containing 10^{12}-10^{20} mutants. However, current genetic transformation tools limit bioengineers to making much smaller mutant libraries. This is relevant in applications where continuous transformation is desired such as in the creation of a library of mutants for drug discovery, metabolic engineering, and deep mutational scanning [1, 2]. In this presentation we will show recent efforts using a continuous flow microfluidic electroporation system to accelerate discovery in protein engineering.

Electroporation is a genetic engineering tool that uses pulsed electric fields for intracellular delivery of molecules such as drugs, proteins, DNA, or RNA and is traditionally performed in a single-batch cuvette system. A subset of the authors have developed a novel microfluidic electroporation platform for improving bacterial electrotransformation efficiency and throughput by up to 3 orders of magnitude [3]. The microfluidic system uses a non-uniform constriction to expose bacteria to a hydrodynamically controlled electric field that induces electroporation. The short residence time (~ms) within the constriction and fast flow rates employed (~mL/min) enables electrotransformation of sufficient bacteria to approach the required numbers needed for protein engineering.

In this study our efforts revolve around re-engineering bacterial two-component sensors, modulating the specificity of bacteriaâ□™s largest class of natural bio-sensors. We have designed a 7kb DNA construct that codes for Spectinomycin resistance and has a theoretical library size of $1x10^{16}$ mutants. *E. coli* cells in exponential growth phase (OD_{600}Â = 0.5) were harvested, washed 3x in 4Â°C 10% glycerol, and concentrated 200x prior to electrotransformation. The 1Â mL samples were transformed in a diverging microfluidic channels at 1.0Â mL/min, 1.5Â mL/min, and 2.0Â mL/min with processing times of 60Â s, 45Â s, or 30Â s, respectively. Preliminary quantification resulted in $5x10^{8}$ (1.0Â mL/min), $2.5x10^{9}$ (1.5Â mL/min), and $3.75x10^{9}$ (2.0Â mL/min) total transformants in comparison to the $2x10^{7}$ total transformants from a 50 Î¼L cuvette transformation. These results suggest that using a single microfluidic channel results in up to a 10x increase in transformation efficiency with an approximately 10-100x increase in throughput compared to cuvette based electroporation.

We envision this microfluidic platform as a tool to enable synthetic biologists to access the largest natural biosensors set in engineered systems. This study focuses on *E. coli* but could be easily applied to other microorganisms and adapted to parallel processing in order to get closer to the theoretical library size. Results of this study will broaden the complexity of engineered libraries that can be developed for protein engineering.

1. Smanski, M.J., et al., *Functional optimization of gene clusters by combinatorial design and assembly.* Nat Biotech, 2014. **32**(12): p. 1241-1249.

2. Fowler, D.M. and S. Fields, *Deep mutational scanning: a new style of protein science.* Nat Meth, 2014. **11**(8): p. 801-807.

3. Garcia, P.A., et al., *High efficiency hydrodynamic bacterial electrotransformation.* Lab Chip, 2017. **17**(3): p. 490-500.

Single-Molecule Analysis of Synthetic RNA Devices in Living Cells By Fluorescence Cross-Correlation Spectroscopy

- Conference: Synthetic Biology Engineering Evolution Design SEED
- Year: 2017
- Proceeding: 2017 Synthetic Biology: Engineering, Evolution & Design (SEED)
- Group: Poster Session
- Session:

  Confirmed Posters

Authors:
Hayden, E.
Ferguson, M.
Roberts, J.
Redman, N.

Bacteria lack compartmentalization and therefore rely on subcellular organization to temporally and spatially localize major cellular processes. As one of the major structures in the cell, the nucleoid possesses positional cues that help direct important functions such as DNA replication and cell division. While some of the factors that contribute to maintaining subcellular organization are characterized, many of the underlying determinants are still unknown. In *Bacillus subtilis*, there remains a class of putative DNA-binding proteins currently uncharacterized, and I hypothesize that some of these proteins use DNA to localize functions important in maintaining subcellular organization. Since many uncharacterized genes lack easily discernable knockout phenotypes, I utilized a gene misexpression approach to identify DNA-binding proteins capable of perturbing some aspect of cellular organization. The misexpression library employed comprises over 800 strains, each containing an uncharacterized gene under the control of an inducible promoter on the *B. subtilis* chromosome. Genes for 98 putative DNA-binding proteins were misexpressed and analyzed for their ability to grow in the presence of inducer. Of these strains, nine exhibited reduced growth and 17 strains did not grow. When analyzed by fluorescence microscopy, 19 of these strains could be associated with a distinct phenotype, including defects in the structure and placement of the chromosome, abnormal cell division, and loss of cell shape. To further characterize strains with the most severe phenotypes we performed RNA-seq experiments to determine if gene misexpression alters the transcription profile. We identified four genes that do not affect transcription, suggesting that the misexpression phenotype is not due to transcriptional regulation. We hypothesize that this class of DNA-binding proteins act as positional cues that help maintain subcellular organization.

Spatial Patterning of Microbial Consortia in Microfluidic Devices

- Conference: Synthetic Biology Engineering Evolution Design SEED
- Year: 2017
- Proceeding: 2017 Synthetic Biology: Engineering, Evolution & Design (SEED)
- Group: Poster Session
- Session:

  Confirmed Posters

Authors:
Alnahhas, R.
Hirning, A., Rice University
Bennett, M. R., Rice University

Small molecule-inducible transcription factors, or sensors, enable temporal control of toxic genes, provide the inputs forÂ *in vivo*Â computation, and allow a host to sense and respond to its environment. In this ongoing work we endeavor to improve the characteristics of several sensors using a mixture of rational engineering and directed evolution. We have improved the dynamic range, sensitivity, and/or specificity of 18 sensors. As an example of the utility of these sensors, we have used five optimized sensors to independently control the expression of five components of a metabolic pathway. By varying the amount of each inducer, we can easily and rapidly optimize the expression profile of the pathway without the need to construct large libraries. To facilitate the adoption of multiplex, sensor-based genetic control, we have created a stable cell-line capable of responding to 11 inputs, thus enabling the modular control of multiple genetic outputs with minimal optimization.

Steps Towards CRISPR-Based Process Control in a Cell-Free System

- Conference: Synthetic Biology Engineering Evolution Design SEED
- Year: 2017
- Proceeding: 2017 Synthetic Biology: Engineering, Evolution & Design (SEED)
- Group: Poster Session
- Session:

  Confirmed Posters

Authors:
Maxwell, C., North Carolina State University
Beisel, C. L., North Carolina State University
Lucks, J. B., Cornell University
Noireaux, V., University of Minnesota

Cell surface glycans make up an important mode of intercellular communication and the enzymes that assemble and degrade them are not only important targets for therapeutics, but also powerful tools for chemical glycobiology. Our recent work has explored 1) the use of engineered glycoside hydrolases for generating universal blood by antigen removal, 2) the identification of inhibitors to glycosyltransferases in cancer pathways, and 3) the development of engineered â□œglycosynthasesâ□□ to assemble and probe the function of glycosaminoglycan structures.

We have previously demonstrated that blood antigen-cleaving glycoside hydrolases can be engineered as tools towards the generation of universal blood through the cleavage of carbohydrate blood group antigens. Identification and directed evolution of these enzymes is enabled by a strategy that involves high-throughput screening with a sensitive fluorescence based assay. Using a similar approach, creatively repurposed to assay for glycosyltransferase rather than glycoside hydrolase activity, we have developed a high-throughput screen for inhibitors targeting fucosyltransferase enzymes that are highly expressed in cancers. Such inhibitors have potential as therapeutics for certain cancers wherein these enzymesâ□™ upregulated activity results in the assembly of carbohydrate structures (e.g. sialyl Lewis^{X}) involved in malignant processes including metastasis. In another focus, we are developing tools for chemoenzymatic glycan synthesis through mechanism-based engineering of â□œglycosynthaseâ□□ enzyme biocatalysts to assemble specific glycosaminoglycan structures. A facile enzymatic synthesis of these glycans will enable glycobiologists to probe their interactions in processes such as neural development and cancer.

Structural Basis for the Enantioselectivity of Est-Y29 Toward (S)-Ketoprofen

- Conference: Synthetic Biology Engineering Evolution Design SEED
- Year: 2017
- Proceeding: 2017 Synthetic Biology: Engineering, Evolution & Design (SEED)
- Group: Poster Session
- Session:

  Confirmed Posters

Authors:
Oh, C.
Ngo, T. D., University of Texas Southwestern Medical Center
Kim, K. K., Sungkyunkwan University School of Medicine

Small molecule-inducible transcription factors, or sensors, enable temporal control of toxic genes, provide the inputs forÂ *in vivo*Â computation, and allow a host to sense and respond to its environment. In this ongoing work we endeavor to improve the characteristics of several sensors using a mixture of rational engineering and directed evolution. We have improved the dynamic range, sensitivity, and/or specificity of 18 sensors. As an example of the utility of these sensors, we have used five optimized sensors to independently control the expression of five components of a metabolic pathway. By varying the amount of each inducer, we can easily and rapidly optimize the expression profile of the pathway without the need to construct large libraries. To facilitate the adoption of multiplex, sensor-based genetic control, we have created a stable cell-line capable of responding to 11 inputs, thus enabling the modular control of multiple genetic outputs with minimal optimization.

SWEET Biosensors for Sugar Transport

- Conference: Synthetic Biology Engineering Evolution Design SEED
- Year: 2017
- Proceeding: 2017 Synthetic Biology: Engineering, Evolution & Design (SEED)
- Group: Poster Session
- Session:

  Confirmed Posters

Authors:
Cheung, L., Carnegie Institution for Science
Frommer, W., Carnegie Institution for Science
The adoptive transfer of T cells expressing chimeric antigen receptors (CARs) has demonstrated clinical efficacy in the treatment of advanced cancers, with anti-CD19 CAR-T cells achieving up to 90% complete remission among patients with relapsed B-cell malignancies. However, challenges such as antigen escape and immunosuppression limit the long-term efficacy of adoptive T-cell therapy. Here, I will discuss the development of next-generation T cells that can target multiple cancer antigens and resist immunosuppression, thereby increasing the robustness of therapeutic T cells against tumor defense mechanisms. This presentation will highlight the potential of synthetic biology in generating novel mammalian cell systems with multifunctional outputs for therapeutic applications.

Synthetic analysis of natural variants yields insights into the evolution and function of auxin signaling F-box proteins in Arabidopsis thaliana

- Conference: Synthetic Biology Engineering Evolution Design SEED
- Year: 2017
- Proceeding: 2017 Synthetic Biology: Engineering, Evolution & Design (SEED)
- Group: Poster Session
- Session:

  Confirmed Posters

Authors:
Wright, R. C., University of Washington
Zahler, M. L., University of Washington
Gerben, S. R., University of Washington
Nemhauser, J., University of Washington

Bacteria lack compartmentalization and therefore rely on subcellular organization to temporally and spatially localize major cellular processes. As one of the major structures in the cell, the nucleoid possesses positional cues that help direct important functions such as DNA replication and cell division. While some of the factors that contribute to maintaining subcellular organization are characterized, many of the underlying determinants are still unknown. In *Bacillus subtilis*, there remains a class of putative DNA-binding proteins currently uncharacterized, and I hypothesize that some of these proteins use DNA to localize functions important in maintaining subcellular organization. Since many uncharacterized genes lack easily discernable knockout phenotypes, I utilized a gene misexpression approach to identify DNA-binding proteins capable of perturbing some aspect of cellular organization. The misexpression library employed comprises over 800 strains, each containing an uncharacterized gene under the control of an inducible promoter on the *B. subtilis* chromosome. Genes for 98 putative DNA-binding proteins were misexpressed and analyzed for their ability to grow in the presence of inducer. Of these strains, nine exhibited reduced growth and 17 strains did not grow. When analyzed by fluorescence microscopy, 19 of these strains could be associated with a distinct phenotype, including defects in the structure and placement of the chromosome, abnormal cell division, and loss of cell shape. To further characterize strains with the most severe phenotypes we performed RNA-seq experiments to determine if gene misexpression alters the transcription profile. We identified four genes that do not affect transcription, suggesting that the misexpression phenotype is not due to transcriptional regulation. We hypothesize that this class of DNA-binding proteins act as positional cues that help maintain subcellular organization.

Synthetic Biology Approaches for the Rational Engineering of Novel E. coli and B. Subtilis Biocatalysts for the Conversion of Industrial Bio-Waste into Biofuels and Bioproducts

- Conference: Synthetic Biology Engineering Evolution Design SEED
- Year: 2017
- Proceeding: 2017 Synthetic Biology: Engineering, Evolution & Design (SEED)
- Group: Poster Session
- Session:

  Confirmed Posters

Authors:
Pasotti, L., University of Pavia
Massaiu, I., University of Pavia
Casanova, M., University of Pavia
Bellato, M., University of Pavia
Magni, P., University of Pavia
Cusella, G., University of Pavia
Calvio, C.
Mazzini, G.
Zucca, S., University of Pavia

Rational engineering of living systems will enable the construction of novel, fully predictable, biological functions to provide solutions in different application fields, like medicine and bioenergy. Synthetic biology aims to provide foundations and appropriate computational/genetic tools to support this task.

Synthetic Biology Memory Circuits in Non-Traditional Hosts

- Conference: Synthetic Biology Engineering Evolution Design SEED
- Year: 2017
- Proceeding: 2017 Synthetic Biology: Engineering, Evolution & Design (SEED)
- Group: Poster Session
- Session:

  Confirmed Posters

Authors:
Khan, N. E., North Carolina Agricultural and Technical State University
Bernstein, H. C., Pacific Northwest National Laboratory
Yeung, E., Caltech
Bonheyo, G., Pacific Northwest National Lab

CRISPR/Cas9 has wide-reaching scientific and therapeutic applications, but efficient, exact targeting of cutting sites can be challenging. Several cas9 variants have been engineered to target alternate PAMs; however, generation of new variants is often laborious. Additionally, the cutting ability of cas9 nucleases is often highly guide-dependent and shows a large amount of variation between sites. In order to engineer new nuclease variants targeted to cleave a given site, we developed a directed evolution platform to select for DNA cleavage in bacteria. Our system allows rapid generation of mutant libraries, which are challenged with phage containing target sites in a competitive pool. Using this method, we developed *S. aureus* cas9 variants which have altered PAM preferences to target cleavage to novel targets of therapeutic relevance.

Synthetic Decoupling of Transcription and Translation Processes for the Quality Control of Gene Expression in Escherichia coli

- Conference: Synthetic Biology Engineering Evolution Design SEED
- Year: 2017
- Proceeding: 2017 Synthetic Biology: Engineering, Evolution & Design (SEED)
- Group: Poster Session
- Session:

  Confirmed Posters

Authors:
Seo, S. W., Seoul National University
Yang, J., Seoul National University
The adoptive transfer of T cells expressing chimeric antigen receptors (CARs) has demonstrated clinical efficacy in the treatment of advanced cancers, with anti-CD19 CAR-T cells achieving up to 90% complete remission among patients with relapsed B-cell malignancies. However, challenges such as antigen escape and immunosuppression limit the long-term efficacy of adoptive T-cell therapy. Here, I will discuss the development of next-generation T cells that can target multiple cancer antigens and resist immunosuppression, thereby increasing the robustness of therapeutic T cells against tumor defense mechanisms. This presentation will highlight the potential of synthetic biology in generating novel mammalian cell systems with multifunctional outputs for therapeutic applications.

Synthetic Gene Networks for 2D Fractal Formation By S. Cerevisiae

- Conference: Synthetic Biology Engineering Evolution Design SEED
- Year: 2017
- Proceeding: 2017 Synthetic Biology: Engineering, Evolution & Design (SEED)
- Group: Poster Session
- Session:

  Confirmed Posters

Authors:
Pothoulakis, G., Imperial College London
Ellis, T., Imperial College London

Multicellular organisms form patterns to enable new functions and aid in specialization. Here, we link the natural capacity for multicellular growth inherent in *Saccharomyces cerevisiae* yeast to externally-controlled synthetic gene networks. In response to IPTG and ATc inductions, our engineered S288C yeast can be made to switch between a unicellular lifestyle and the multicellular filamenting patterns usually seen with pseudohyphal growth. Fine-control of dynamic behavior and gene expression within a colony of these cells provides us with the framework for genetic encoding of structural forms, with 2D fractal-like patterns being our goal. We show here how changes in the expression of the *BUD* gene family can alter the pattern of growth and we link our phenotype switches to genetically-encoded timer circuits in order control the duration that a colony grows either in multicellular or unicellular form. To further our ability to define the 2D fractal we also developed new tools to aid differentiation as a colony grows. We made and characterized new hybrid externally-inducible promoters that only express in mother cells after they have budded a daughter cell. These were designed in order to control how often filaments form new branches in our fractal growth phenotype. These new mother-specific promoters provide a unique type of differentiated regulation that could be applied to other synthetic biology designs where cells within the population diverge to perform alternative tasks.

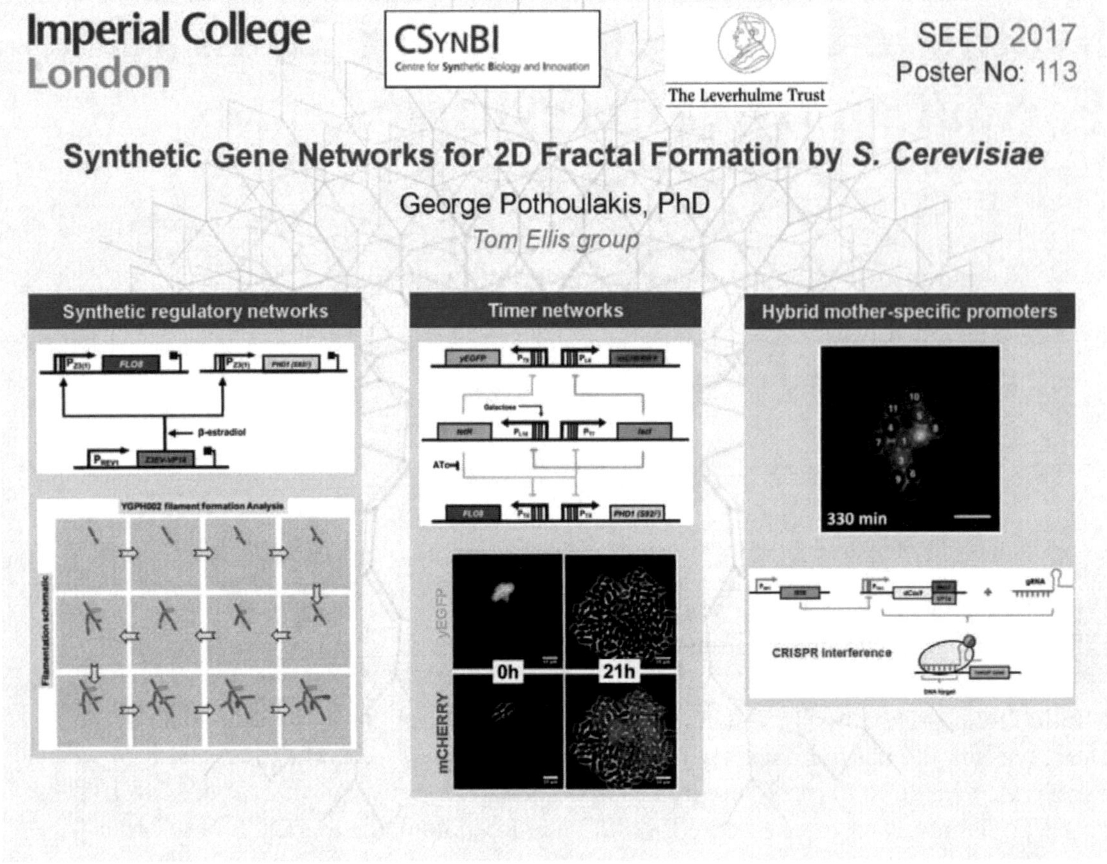
Imperial College London
CSynBI
Centre for Synthetic Biology and Innovation
The Leverhulme Trust
SEED 2017
Poster No: 113
Synthetic Gene Networks for 2D Fractal Formation by S. Cerevisiae
George Pothoulakis, PhD
Tom Ellis group
Synthetic regulatory networks
FLO8
PHD1 (S937)
β-estradiol
Z3EV-VP16
Filamentation schematic
Timer networks
yEGFP
Galactose
tetR
lacI
FLO8
PHD1 (S937)
yEGFP
mCHERRY
0h
21h
Hybrid mother-specific promoters
330 min
dCas9
gRNA
CRISPR Interference

Synthetic Morphogenesis in Plant Roots

- Conference: Synthetic Biology Engineering Evolution Design SEED
- Year: 2017
- Proceeding: 2017 Synthetic Biology: Engineering, Evolution & Design (SEED)
- Group: Poster Session
- Session:

  Confirmed Posters

Authors:
Brophy, J. A. N., Carnegie Institution of Washington
Dinneny, J. R., Carnegie Institution of Washington
Protein engineering has important applications that range from developing sensors in bacteria to producing antibodies for cancer therapeutics. Re-engineering protein function usually requires many amino acid changes, which results in libraries containing 10^{12}-10^{20} mutants. However, current genetic transformation tools limit bioengineers to making much smaller mutant libraries. This is relevant in applications where continuous transformation is desired such as in the creation of a library of mutants for drug discovery, metabolic engineering, and deep mutational scanning [1, 2]. In this presentation we will show recent efforts using a continuous flow microfluidic electroporation system to accelerate discovery in protein engineering.

Electroporation is a genetic engineering tool that uses pulsed electric fields for intracellular delivery of molecules such as drugs, proteins, DNA, or RNA and is traditionally performed in a single-batch cuvette system. A subset of the authors have developed a novel microfluidic electroporation platform for improving bacterial electrotransformation efficiency and throughput by up to 3 orders of magnitude [3]. The microfluidic system uses a non-uniform constriction to expose bacteria to a hydrodynamically controlled electric field that induces electroporation. The short residence time (~ms) within the constriction and fast flow rates employed (~mL/min) enables electrotransformation of sufficient bacteria to approach the required numbers needed for protein engineering.

In this study our efforts revolve around re-engineering bacterial two-component sensors, modulating the specificity of bacteriaâ□™s largest class of natural bio-sensors. We have designed a 7kb DNA construct that codes for Spectinomycin resistance and has a theoretical library size of $1x10^{16}$ mutants. *E. coli* cells in exponential growth phase (OD_{600}Â = 0.5) were harvested, washed 3x in 4Â°C 10% glycerol, and concentrated 200x prior to electrotransformation. The 1Â mL samples were transformed in a diverging microfluidic channels at 1.0Â mL/min, 1.5Â mL/min, and 2.0Â mL/min with processing times of 60Â s, 45Â s, or 30Â s, respectively. Preliminary quantification resulted in $5x10^8$ (1.0Â mL/min), $2.5x10^9$ (1.5Â mL/min), and $3.75x10^9$ (2.0Â mL/min) total transformants in comparison to the $2x10^7$ total transformants from a 50 Î¼L cuvette transformation. These results suggest that using a single microfluidic channel results in up to a 10x increase in transformation efficiency with an approximately 10-100x increase in throughput compared to cuvette based electroporation.

We envision this microfluidic platform as a tool to enable synthetic biologists to access the largest natural biosensors set in engineered systems. This study focuses on *E. coli* but could be easily applied to other microorganisms and adapted to parallel processing in order to get closer to the theoretical library size. Results of this study will broaden the complexity of engineered libraries that can be developed for protein engineering.

1. Smanski, M.J., et al., *Functional optimization of gene clusters by combinatorial design and assembly.* Nat Biotech, 2014. **32**(12): p. 1241-1249.

2. Fowler, D.M. and S. Fields, *Deep mutational scanning: a new style of protein science.* Nat Meth, 2014. **11**(8): p. 801-807.

3. Garcia, P.A., et al., *High efficiency hydrodynamic bacterial electrotransformation.* Lab Chip, 2017. **17**(3): p. 490-500.

Synthetic Sugar Cassettes for the Efficient Production of Flavonol Glycosides in Escherichia coli

- Conference: Synthetic Biology Engineering Evolution Design SEED
- Year: 2017
- Proceeding: 2017 Synthetic Biology: Engineering, Evolution & Design (SEED)
- Group: Poster Session
- Session:

  Confirmed Posters

Authors:
Parajuli, P., Sun Moon University
Pandey, R. P., Sun Moon University
Sohng, J. K., Sun Moon University

Our ability to design cell factories to produce valuable chemicals requires the â□œrecombinationâ□□ not only of existing but also designer enzymes into novel metabolic pathways to achieve entirely new metabolic function. This poses two distinct challenges that need to be solved in an integrated way if we ambition to fully deliver on the promise of synthetic biology. The first challenge deals with the design, at scale, of novel enzymes with high level of activities for reactions not known to be catalyzed in nature. To this end, Arzeda has developed and scaled high-throughput computational methodologies to exploit natural enzyme latent catalytic promiscuity and rapidly design new biocatalysts. A successful industrial application of this technology will be presented.

The second challenge is that of finding optimal ways to arrange natural and designed enzymes to biosynthetize a small-molecule of interest. To illustrate our progress towards solving this challenge, we will discuss a novel computational tool that Arzeda has developed for the automated design of novel biosynthetic pathways. Inspired by retrosynthetic methods in organic chemistry, our software draws on databases of known natural enzymatic reactions as well as reactions that can be catalyzed by designed enzymes to exhaustively enumerate biosynthetic routes from a set of desired metabolites down to any small molecule of interest. Pathways are ranked based on thermodynamic feasibility and other metrics. We will discuss applications of industrial relevance in the field of fine and bulk chemicals.

Syntrack 2: A Scalable, Automated DNA Assembly Production Workflow Management

- Conference: Synthetic Biology Engineering Evolution Design SEED
- Year: 2017
- Proceeding: 2017 Synthetic Biology: Engineering, Evolution & Design (SEED)
- Group: Poster Session
- Session:

  Confirmed Posters

Authors:
Meng, J., DOE Joint Genome Institute, Lawrence Berkeley National Laboratory
Tarver, A., DOE Joint Genome Institute, Lawrence Berkeley National Laboratory
Hamilton, M., DOE Joint Genome Institute, Lawrence Berkeley National Laboratory
Evans, R., DOE Joint Genome Institute
Hillson, N. J., Joint BioEnergy Institute
Cheng, J. F., Lawrence Berkeley National Laboratory
Deutsch, S., Lawrence Berkeley National Laboratory
Simirenko, L., DOE Joint Genome Institute, Lawrence Berkeley National Laboratory
Mosquito-borne diseases such as Zika, dengue fever and Chikungunya are major and increasing international public health concerns. The primary vector of these viral diseases is *Aedes aegypti* and current control measures against this mosquito are proving inadequate or problematic. An innovative and effective control technique for *Aedes aegypti* involves the release of self-limiting mosquitoes. These mosquitoes are genetically engineered to be homozygous for a self-limiting gene under control of an on/off tetracycline switch. When released into the wild, any offspring of the transgenic males will inherit one copy of the transgene. In the absence of tetracycline antidote, the transgene will prevent the mosquito from developing to adulthood, thus suppressing the population with repeated releases. Oxitecâ□™s self-limiting transgenic insect technology has also been applied successfully to develop population control measures for major crop pests such as the Mediterranean fruit fly (Medfly) and the diamondback moth. In a variation on the self-limiting genetics used in *Aedes* mosquitoes, genetic sexing strains that express the self-limiting gene in a sex-specific manner are used to produce male-only cohorts of these crop pests for release in control programmes. These and other developments in Oxitecâ□™s genetic engineering programme to enable biological control of major insect pests will be discussed. The Oxitec transgenic mosquito method has dramatically reduced *Aedes aegypti* populations in trials around the world and is now being used under operational conditions in the Cayman Islands and Brazil to control *Aedes aegypti*. Results of these programmes will be presented.

Systems and Synthetic Biology of Rhodococcus Opacus to Enable Conversion of Lignin-Derived Aromatic Compounds into Lipids

- Conference: Synthetic Biology Engineering Evolution Design SEED
- Year: 2017
- Proceeding: 2017 Synthetic Biology: Engineering, Evolution & Design (SEED)
- Group: Poster Session
- Session:

  Confirmed Posters

Authors:
Moon, T. S., Washington University in St. Louis
Henson, W. R., Washington University in St. Louis
DeLorenzo, D., Washington University in St. Louis
Rottinghaus, A., Washington University in St. Louis

As synthetic biology transitions out of the lab and towards real-world applications, there is a need to develop a well-characterized toolbox of genetic sensors and regulators that can respond to extrinsic environmental conditions, instead of lab chemicals (e.g., IPTG). In addition, many sophisticated cellular functions require complex genetic circuits, which can lead to significant consumption of finite cellular resources and thus resource-coupled interference in predictable circuit behavior. To solve these problems, we have developed environmental sensors, RNA regulators, and feedback circuits. First, in multiple bacterial species, we have built and characterized sensors that respond to pH, temperature, light, oxygen levels, nitrogen source levels, or concentrations of diverse compounds found in a variety of environments. These environmental sensors have also been combined to construct multi-input logic gates (e.g., AND gates). Second, we have employed the CRISPR interference system (CRISPRi) and synthetic antisense RNAs (asRNAs) to repress or derepress target genes in a programmable manner. Importantly, we provided design rules for asRNA regulators, based on a comprehensive study of more than a hundred *de novo* designed asRNA regulators, enabling predictable implementation of these regulators in complex genetic circuits (e.g., NAND gates that respond to pH and temperature). Third, we have investigated the effects of resource competition on circuit behaviors. For the first time, we experimentally demonstrated that negative feedback can significantly reduce resource-coupled interference in complex circuits, allowing for robust circuit performance. These environmental sensors, RNA regulators, and feedback motif lay the groundwork for the implementation of complex, environmentally-responsive genetic circuits in metabolic engineering, environmental, and medical applications.

Tackling Maple Syrup Urine Disease (MSUD) with Engineered Microbes

- Conference: Synthetic Biology Engineering Evolution Design SEED
- Year: 2017
- Proceeding: 2017 Synthetic Biology: Engineering, Evolution & Design (SEED)
- Group: Poster Session
- Session:

  Confirmed Posters

Authors:
LI, N., Synlogic Inc.
Millet, Y., Synlogic Inc.
Isabella, V. M., Synlogic Inc.
Ha, B., Synlogic Inc.
Walsh, K., Synlogic Inc
Li, K., Synlogic, Inc.
West, K., Synlogic Inc.
Castillo, M., Synlogic Inc.
Falb, D., Synlogic, Inc.
Miller, P., Synlogic Inc.
We have engineered *Escherichia coli* to overproduce aliphatic methyl ketones (MK) in the C_{11} to C_{15} (diesel) range; this group of MK includes 2-undecanone and 2-tridecanone, which have favorable cetane numbers and are also of importance to the flavor and fragrance industry. Overall, we have made specific improvements that resulted in 50,000-fold enhancement in MK titer relative to that of a fatty acid-overproducing *E. coli* strain. The first generation of engineered *E. coli* (Goh et al. 2012) had modifications that included: (a) overproduction of Î²-ketoacyl-coenzyme A (CoA) thioesters achieved by modification of the Î²-oxidation pathway (specifically, overexpression of a heterologous acyl-CoA oxidase and native FadB, and chromosomal deletion of *fadA*) and (b) overexpression of a native thioesterase (FadM). Subsequent genetic modifications included balancing overexpression of *fadR* and *fadD* to increase fatty acid flux into the pathway, consolidation of the pathway from two plasmids into one, codon optimization, and knocking out key acetate production pathways (resulting in strain EGS1895). This strain produced an MK titer of 1.4 g/L with 1% glucose in shake-flask experiments, representing 40% of the maximum theoretical yield. ^{13}C-based metabolic flux analysis with EGS1895 showed increased carbon flux through the pentose phosphate pathway (PPP), which we hypothesized is to generate more NADPH - an important co-factor for FabG (Î²-ketoacyl-ACP reductase) activity during fatty acid biosynthesis.Â In an attempt to divert flux away from PPP, we overexpressed alternate FabG variants that we have demonstrated preferentially use NADH instead of NADPH. Specifically, overexpression

of the NADH-dependent *Acholeplasma laidlawii* FabG increased MK production by 20-50% in shake-flask experiments and had more sustained MK production in fed-batch fermentation compared to EGS1895. We have achieved MK titers of ~6.0 g/L in fed-batch fermentation, which is the best MK titer observed to date.

The Application of miRNA in Bone Tissue Engineering

- Conference: Synthetic Biology Engineering Evolution Design SEED
- Year: 2017
- Proceeding: 2017 Synthetic Biology: Engineering, Evolution & Design (SEED)
- Group: Poster Session
- Session:

  Confirmed Posters

Authors:
Chew, S. A.
Arriaga, M. A.
Gutierrez, A.

Bacteria lack compartmentalization and therefore rely on subcellular organization to temporally and spatially localize major cellular processes. As one of the major structures in the cell, the nucleoid possesses positional cues that help direct important functions such as DNA replication and cell division. While some of the factors that contribute to maintaining subcellular organization are characterized, many of the underlying determinants are still unknown. In *Bacillus subtilis*, there remains a class of putative DNA-binding proteins currently uncharacterized, and I hypothesize that some of these proteins use DNA to localize functions important in maintaining subcellular organization. Since many uncharacterized genes lack easily discernable knockout phenotypes, I utilized a gene misexpression approach to identify DNA-binding proteins capable of perturbing some aspect of cellular organization. The misexpression library employed comprises over 800 strains, each containing an uncharacterized gene under the control of an inducible promoter on the *B. subtilis* chromosome. Genes for 98 putative DNA-binding proteins were misexpressed and analyzed for their ability to grow in the presence of inducer. Of these strains, nine exhibited reduced growth and 17 strains did not grow. When analyzed by fluorescence microscopy, 19 of these strains could be associated with a distinct phenotype, including defects in the structure and placement of the chromosome, abnormal cell division, and loss of cell shape. To further characterize strains with the most severe phenotypes we performed RNA-seq experiments to determine if gene misexpression alters the transcription profile. We identified four genes that do not affect transcription, suggesting that the misexpression phenotype is not due to transcriptional regulation. We hypothesize that this class of DNA-binding proteins act as positional cues that help maintain subcellular organization.

The Genetic Incompatibility of Metallo-?-Lactamases: Synthetic Biology Lessons from Directed Evolution

- Conference: Synthetic Biology Engineering Evolution Design SEED
- Year: 2017
- Proceeding: 2017 Synthetic Biology: Engineering, Evolution & Design (SEED)
- Group: Poster Session
- Session:

  Confirmed Posters

Authors:
Socha, R., University of British Columbia
Tokuriki, N., University of British Columbia

State-of-the-art strain engineering techniques for the protein producing yeast host *Pichia pastoris* include overexpression of homologous and heterologous genes, and deletion of host genes. For this purpose overexpression vectors and gene deletion methods such as the split marker technique have been established. For metabolic and cell engineering purposes, the simultaneous overexpression of more than one gene is often needed. Previous approaches employing subsequent steps of overexpression and marker recycling were time- and labor-consuming. Therefore, efficient systems allowing multiple gene overexpression are required, that can be stably integrated into the *P. pastoris* genome. To this end, we developed a synthetic biology toolbox based on Golden Gate cloning to enable efficient construction of complex and versatile over-expression vectors. Up to five different expression cassettes, employing a library of promoters and terminators can be combined into one vector, and successfully integrated into the genomic DNA of *P. pastoris* at targeted loci in one step. Recent trends in synthetic biology, however, go into the direction of building up large and complex reaction networks. To allow for clean and unscarred genetic engineering, a CRISPR/Cas9 based method for gene insertions, deletions and replacements was developed, which paves the way for precise genomic rearrangements in *P. pastoris*. By using this technique precise genomic integrations were performed efficiently without integrative selection markers. The repertoire of genetic techniques developed so far, will provide a wide variety of possibilities to engineer *P. pastoris*. Applications for these synthetic biology tools in cell engineering of recombinant *P. pastoris* will be presented.

The Sweet Smell of Biocatalysis: Using Synthetic Biochemistry for the in Vitro production of Biofuels and Commodity Chemicals

- Conference: Synthetic Biology Engineering Evolution Design SEED
- Year: 2017
- Proceeding: 2017 Synthetic Biology: Engineering, Evolution & Design (SEED)
- Group: Poster Session
- Session:

  Confirmed Posters

Authors:
Korman, T., UCLA

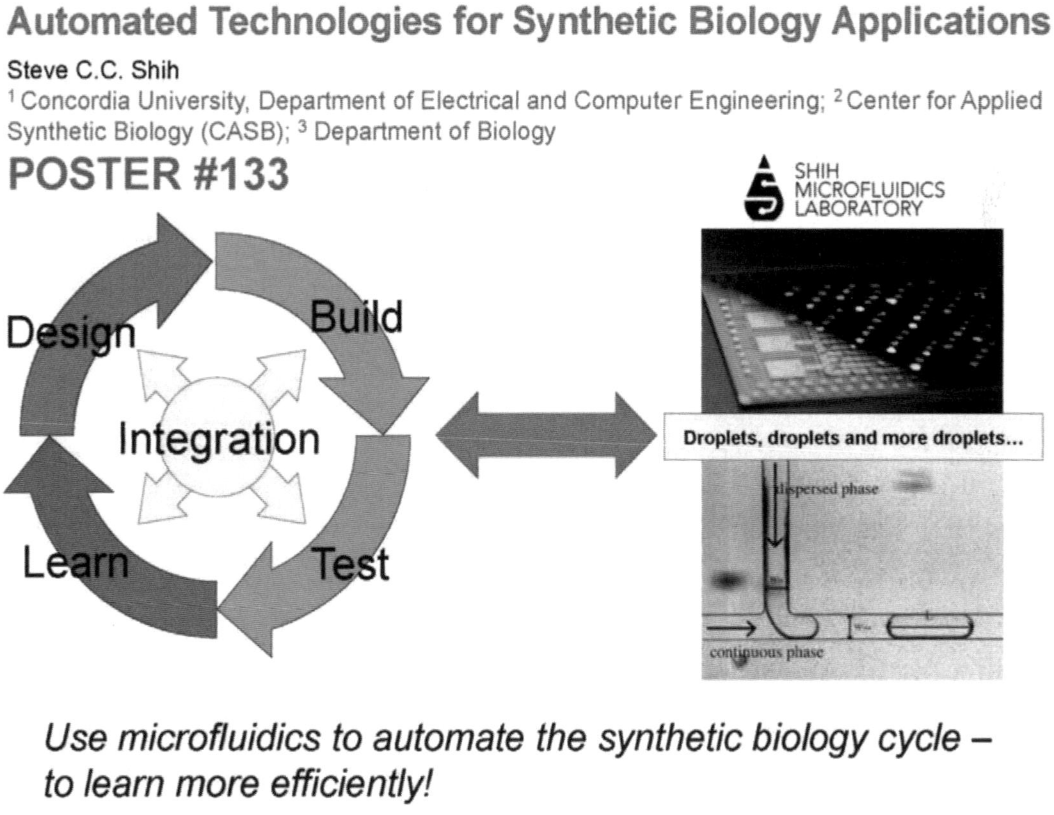

Towards a Cell-Free System for the Discovery of Post-Translationally Modified Peptide Natural Products

- Conference: Synthetic Biology Engineering Evolution Design SEED
- Year: 2017
- Proceeding: 2017 Synthetic Biology: Engineering, Evolution & Design (SEED)
- Group: Poster Session
- Session:

  Confirmed Posters

Authors:
Mohr, B. P., Oak Ridge National Laboratory
Pelletier, D. A., Oak Ridge National Laboratory
Doktycz, M. J., Oak Ridge National Laboratory
Blair, P. M., Oak Ridge National Laboratory

State-of-the-art strain engineering techniques for the protein producing yeast host *Pichia pastoris* include overexpression of homologous and heterologous genes, and deletion of host genes. For this purpose overexpression vectors and gene deletion methods such as the split marker technique have been established. For metabolic and cell engineering purposes, the simultaneous overexpression of more than one gene is often needed. Previous approaches employing subsequent steps of overexpression and marker recycling were time- and labor-consuming. Therefore, efficient systems allowing multiple gene overexpression are required, that can be stably integrated into the *P. pastoris* genome. To this end, we developed a synthetic biology toolbox based on Golden Gate cloning to enable efficient construction of complex and versatile over-expression vectors. Up to five different expression cassettes, employing a library of promoters and terminators can be combined into one vector, and successfully integrated into the genomic DNA of *P. pastoris* at targeted loci in one step. Recent trends in synthetic biology, however, go into the direction of building up large and complex reaction networks. To allow for clean and unscarred genetic engineering, a CRISPR/Cas9 based method for gene insertions, deletions and replacements was developed, which paves the way for precise genomic rearrangements in *P. pastoris*. By using this technique precise genomic integrations were performed efficiently without integrative selection markers. The repertoire of genetic techniques developed so far, will provide a wide variety of possibilities to engineer *P. pastoris*. Applications for these synthetic biology tools in cell engineering of recombinant *P. pastoris* will be presented.

Towards the Automated Design of Bacterial Genomes for Tailored Applications

- Conference: Synthetic Biology Engineering Evolution Design SEED
- Year: 2017
- Proceeding: 2017 Synthetic Biology: Engineering, Evolution & Design (SEED)
- Group: Poster Session
- Session:

  Confirmed Posters

Authors:
Salis, H., Pennsylvania State University

Protein glycosylation, the post-translational attachment of complex oligosaccharides (glycans), is the most abundant polypeptide modification in eukaryotes, and is integrally involved in human biology and disease. Changes in glycosylation have been shown to modulate the pharmacokinetics and potency of protein therapeutics and vaccines, making them critical quality attributes for biotechnology applications. However, the study of glycans in native systems (glycobiology) and the intentional manipulation of protein glycosylation patterns (glycoengineering) remain limited by a shortage of methods to characterize glycosylation enzyme activities and a reliance on endogenous glycosylation systems which can produce highly variable glycosylation patterns and constrain choices of production hosts, protein trafficking, and possible glycoforms. The structural complexity of enzymes involved in initial modification of polypeptides with glycans and the vast sequence spaces which must be surveyed to determine their specificities have impeded efforts to implement synthetic glycosylation systems in heterologous production hosts, such as *Escherichia coli*. New tools are required to rapidly express polypeptide modifying glycosyltransferases and determine design rules for engineering protein glycosylation sites which can be efficiently and specifically modified with a glycan of interest.

Here we describe an *in vitro* platform for high-throughput expression and characterization of glycosylation enzymes using *E. coli* Cell-Free Protein Synthesis (CFPS) and Self-Assembled Monolayers for Desorption Ionization Mass Spectrometry (SAMDI-MS). This workflow allowed us to produce >800 μg/mL of a cytoplasmic N-linked glycosyltransferase (NGT) *in vitro* and determine its peptide acceptor and sugar donor specificities at unprecedented depth and throughput with ~10,000 unique reactions conditions and ~3,000 unique peptide substrates. We used this dataset to develop a small, robust acceptor sequence motif (GlycTag) to direct the efficient installation of N-linked glycans onto the internal loops of heterologous protein substrates *in vitro* and in the cytoplasm of living *E. coli*. We found that SAMDI-MS observations at the peptide level accurately predicted modification trends in the context of whole proteins. We placed GlycTags in three structurally diverse target proteins, including the Fc region of a human antibody, and found that these sequences were modified at 3-5 fold greater efficiency compared to wild-type glycosylation sequence motifs. Notably, this N-linked glycosylation system does not require protein transport across cellular membranes or the use of membrane bound components, making it particularly attractive for the development of synthetic glycosylation systems.

This work provides a disruptive technique for glycosyltransferase characterization and provides a new set of tools towards emerging bacterial and *in vitro* glycosylation platforms. Ultimately, these glycoengineering efforts will enable a deeper understanding of glycan structure and function and help to bring about a new generation of rationally designed glycoprotein therapeutics, vaccines, and diagnostics.

Transcriptional Reprogramming with dCas9 of Saccharomyces Cerevisiae metabolic Pathways

- Conference: Synthetic Biology Engineering Evolution Design SEED
- Year: 2017
- Proceeding: 2017 Synthetic Biology: Engineering, Evolution & Design (SEED)
- Group: Poster Session
- Session:

  Confirmed Posters

Authors:
Damgaard Jensen, E., Technical University of Denmark
Jakociunas, T., The Novo Nordisk Foundation Center for Biosustainability
Jensen, M. K., Technical University of Denmark
Keasling, J. D., Joint BioEnergy Institute (JBEI)

The availability of cheap and scalable gene synthesis techniques will open up the exploration of sequence-function relationships at unprecedented scales. Attempts to reduce the costs of gene synthesis have focused on how to assemble microarray derived oligo (OLS) pools into longer length genes. Here we present a new approach capable of assembling hundreds of genes simultaneously from a complex oligo pool by localizing only the fragments required for each individual gene assembly to barcoded beads before isolating the assembly reaction inside an emulsion droplet. Multiplexed reactions of 384 genes, each assembled from five oligos, have costs approaching $1 per 500 bp gene. We apply this to assemble a thousand phylogenetically diverse orthologs of phosphopantetheine adenylyltransferase (PPAT) representing over 500 kbp of coding sequence and use a sequencing based barcode readout to carry out a complementation assay in E. coli, highlighting how this technique can be applied to determine functional relationships of long sequences in high-throughput at low-cost.

Tuning the Dials of Metabolite Biosensors

- Conference: Synthetic Biology Engineering Evolution Design SEED
- Year: 2017
- Proceeding: 2017 Synthetic Biology: Engineering, Evolution & Design (SEED)
- Group: Poster Session
- Session:

  Confirmed Posters

Authors:
Liu, D.
Oyarzún, D., Imperial College London
Mannan, A., Imperial College London
Zhang, F., Washington University

Genetically-encoded biosensors trigger expression of target genes in response to metabolites or small molecules. They offer great potential for diverse applications such as dynamic pathway engineering, environmental sensing, and real-time metabolic monitoring.Â Biosensors can be fine-tuned by modifying their DNA sequences, yet so far we have a poor grasp of how such tuning shapes the resulting dose-response curve. We built and characterized promoter libraries in E. coli controlled by metabolite-responsive transcription factors, and asked how changes in the operator sequence shape the biosensor's dose-response curve. Our data from the lactose and arginine systems revealed that the biosensors' dynamic range and response threshold are inherently coupled to one another. Mathematical modelling revealed a highly nonlinear relation between dynamic range and threshold, exhibiting maximum achievable limits for both parameters. Combining theory and experiments we uncovered two other tuning dials that provide orthogonal control of the dose-response curve. Our work sheds light on the limits of synthetic biology designs and provides guidelines for biosensor design in metabolic engineering.

* Di Liu and Ahmad Mannan contributed equally to this work.

* **Correspondence:** fzhang@seas.wustl.edu, d.oyarzun@imperial.ac.uk

#79 Tuning the Dials of Transcription Factor-based Metabolite Biosensors

Di Liu[1,†], Ahmad Mannan[2,†], Fuzhong Zhang[1,*], Diego Oyarzún[2,*]

1. Department of Energy, Environmental and Chemical Engineering, Washington University in St. Louis, MO, USA
2. Biomathematical Sciences, Imperial College London, London, UK

† Equal contribution

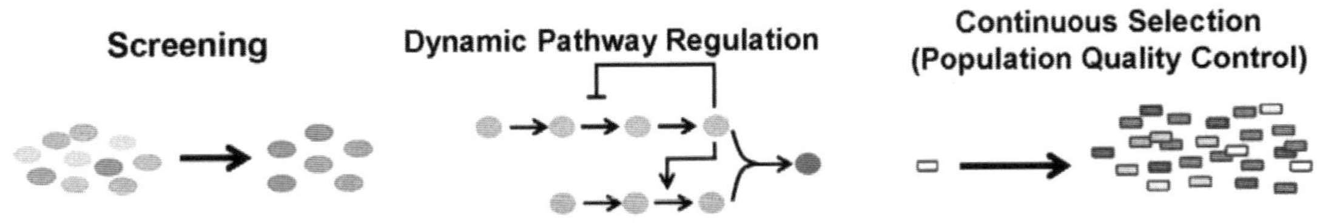

Dynamic Range & Response Threshold

Question: How to tune the response curve of biosensors?

Design Constrain

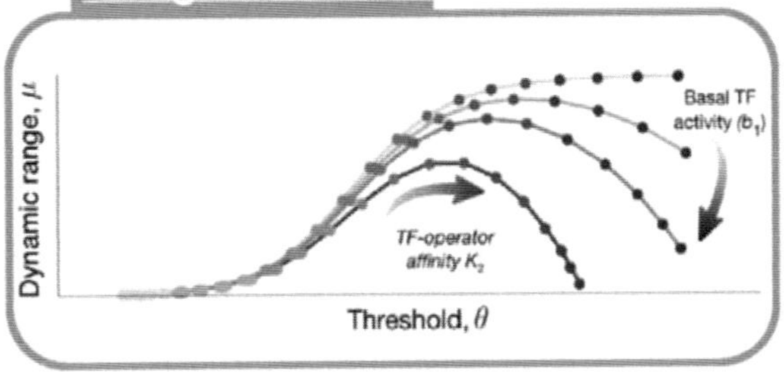

Orthogonal Control

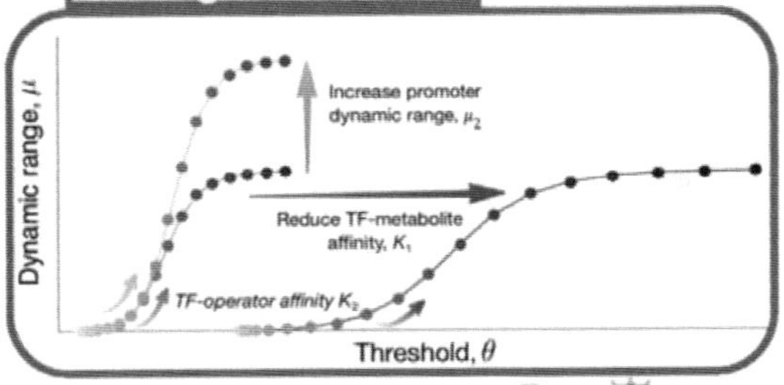

Funding sources

Tuning the Yeast MAPK Pathway for Yeast-Based Biosensors

- Conference: Synthetic Biology Engineering Evolution Design SEED
- Year: 2017
- Proceeding: 2017 Synthetic Biology: Engineering, Evolution & Design (SEED)
- Group: Poster Session
- Session:

  Confirmed Posters

Authors:
Stainbrook, S., Northwestern University
Tyo, K. E. J., Northwestern University

DNA assembly forms the cornerstone of modern synthetic biology. Despite the numerous available methods, scarless multi-fragment assembly of large plasmids remains challenging. Furthermore, the upcoming wave in molecular biological automation demands a rethinking of how we perform DNA assembly. To streamline automation workflow and minimize operator intervention, a non-enzymatic assembly method is highly desirable. Here, we report the optimization and operationalization of a process called Twin-Primer Assembly (TPA), which is a method to assemble PCR-amplified fragments into a plasmid without the use of enzymes.

Each fragment used in TPA requires two PCR products, amplified separately using two sets of primers. One product has an overlap to the fragment that comes before and the other has an overlap to the fragment that comes after. The length of the overlap region is determined by the melting temperature (T_M) of the overlap region, and all overlaps used in an assembly are designed to have the same T_M. The two PCR products are mixed and fully denatured at high temperature before they are allowed to re-anneal slowly, generating intermediates that have overhangs on both sides. These intermediates are then allowed to hybridize at an elevated temperature to form a circular nicked plasmid. Upon *E. coli* transformation, the cells phosphorylate and ligate the nicks to create the final product plasmid.

The TPA non-enzymatic DNA assembly method is scarless and generally sequence independent. It is capable of assembling a 7 kb plasmid from 10 fragments at close to 80% fidelity, and a 31kb plasmid from 5 fragments at around 50% fidelity. Assembling more than 10 fragments is likely possible for constructing smaller plasmids, while assembling a plasmid larger than 31 kb is likely possible using fewer fragments. Even without the use of enzymes, the performance of TPA is on par with some of the best *in vitro* assembly methods currently available. TPA should be an invaluable addition to a synthetic biologistâ□™s toolbox.

Understanding and Manipulating Chromosomal-Scale DNA Looping in Escherichia coli

- Conference: Synthetic Biology Engineering Evolution Design SEED
- Year: 2017
- Proceeding: 2017 Synthetic Biology: Enginecring, Evolution & Design (SEED)
- Group: Poster Session
- Session:

  Confirmed Posters

Authors:
Hao, N., The University of Adelaide
Shearwin, K. E., The University of Adelaide
Dodd, I. B., The University of Adelaide

Our understanding of translation underpins our capacity to engineer living systems. The canonical start codon (AUG) and a few near-cognates (GUG, UUG) are considered as the â□œstart codonsâ□□ for translation initiation in *Escherichia coli*. Translation is typically not thought to initiate from the 61 remaining codons. Here, we quantified translation initiation of green fluorescent protein and nanoluciferase in *E. coli* from all 64 triplet codons and across a range of DNA copy number. We detected initiation of protein synthesis above measurement background for 47 codons. Translation from non-canonical start codons ranged from 0.007% to 3% relative to translation from AUG. Translation from 17 non-AUG codons exceeded the highest reported rates of non-cognate codon recognition. Translation initiation from non-canonical start codons may contribute to the synthesis of peptides in both natural and synthetic biological systems.

The presence of frequent but very low-level expression of proteins via non-canonical start codons would have widespread implications for genome annotation, cellular engineering, and our fundamental understanding of translation initiation. We encourage reconsidering existing definitions and further exploration of what is considered a start codon.

Using Yeast to Identify Peptides That Regulate Human Oncogenic Transcription Factors

- Conference: Synthetic Biology Engineering Evolution Design SEED
- Year: 2017
- Proceeding: 2017 Synthetic Biology: Engineering, Evolution & Design (SEED)
- Group: Poster Session
- Session:

  Confirmed Posters

Authors:
Wu, C. F.
Koschwanez, J.
Plants are a rich source of unique scaffolds, including 25% of natural-product-derived drugs. However, the discovery, synthesis, and overall material supply chains for sourcing plant natural products and their derivatives remain ad hoc, biased, and tedious. While microbial biosynthesis presents compelling alternatives to traditional approaches based on extraction from natural plant hosts, many challenges exist in the reconstruction of plant specialized metabolic pathways in microbial hosts. My laboratory has developed approaches to address the challenges that arise in the reconstruction of complex biosynthesis schemes, including spatial engineering strategies to direct the activities and specificities of pathway enzymes and recoding strategies to address folding, processing, and stability issues that may arise with the expression of plant enzymes in heterologous microbial hosts. We have utilized these strategies to develop yeast-based production platforms for an important class of plant alkaloids, the benzylisoquinoline alkaloids, including the medicinal opioids. These synthetic biology platforms will lead to transformative advances in natural product discovery, drug development, and production.

Vectoreditorjs: An Open-Source Web-Based Javascript Application for DNA Sequence Visualization, Annotation, and Editing

- Conference: Synthetic Biology Engineering Evolution Design SEED
- Year: 2017
- Proceeding: 2017 Synthetic Biology: Engineering, Evolution & Design (SEED)
- Group: Poster Session
- Session:

  Confirmed Posters

Authors:
LaFrance, S., JBEI
Springthorpe, L., Joint Bioenergy Institute
Plahar, H., Joint BioEnergy Institute
Coble, J., Joint BioEnergy Institute, Lawrence Berkeley National Laboratory
Rich, T., TeselaGen Biotechnologies Inc
Hillson, N. J., Joint BioEnergy Institute
Nixon, J., UC Berkeley

Synthetic Biology aims to implement new cellular features through the design of original genetic parts and systems. A wide variety of circuit architectures have been built successfully, proving the ability of synthetic genetic parts to predictably interact with each other within a single bacterial cell [4]. The next stage of synthetic biology will involve the *de novo* engineering of self-organizing multicellular systems. One of the main methods found in nature for creating complex spatially distributed systems is cellular differentiation through asymmetric cell division (ACD), whereby stem cells spawn off irreversibly differentiated daughters. The presence of different cell lineages within an eukaryotic tissue allows for the partitioning of metabolic and structural tasks. Furthermore, if one or both of the daughter cells becomes incapable of dividing, controlling asymmetric division probability can lead to tissue size and shape control [5]. In complex eukaryotic organisms ACD is typically brought about by the differential expression of key genes within each cell type. Here, we engineered a novel method for generating ACD in *Escherichia coli* and regulating its frequency within the population. This work provides a platform for the construction of complex synthetic tissues and organisms from bacterial cells that differentiate during development of the population.

Current methods for creating differentiated cell types in synthetic bacteria utilize gene regulatory networks that allow for bistable gene expression [2, 1]. While such systems generate bimodal populations, they are very sensitive to both intrinsic and extrinsic noise sources that can randomly force transitions between states [3]. We use a novel method for creating terminally differentiated synthetic bacterial cells. Specifically, we refactored DNA shuttling mechanisms from *Caulobacter crescentus*

into *E. coli* to target plasmid DNA into one and only one daughter cell upon division. Thus, the two daughter cells will not just have different gene expression states, but will be irreversibly genetically distinct. Our findings will lead to synthetic "tissues" that can be used in complex settings.

References

[1] R. G. Egbert and E. Klavins. Fine-tuning gene networks using simple sequence repeats.*Proceedings of the National Academy of Sciences*, 109(42):16817â□“16822, 2012.
[2] T. S. Gardner, C. R. Cantor, and J. J. Collins. Construction of a genetic toggle switch in *Escherichia coli. Nature*, 403(6767):339â□“342, 2000.
[3] C. Gupta, J. M. LÃ³pez, W. Ott, K. JosiÄ‡, and M. R. Bennett. Transcriptional delay stabilizes bistable gene networks. *Physical review letters*, 111(5):058104, 2013.
[4] A. S. Khalil and J. J. Collins. Synthetic biology: applications come of age. *Nature Reviews*

Genetics, 11(5):367â□“379, 2010.
[5] A. D. Lander, K. K. Gokoffski, F. Y. Wan, Q. Nie, and A. L. Calof. Cell lineages and the logic of proliferative control. *PLoS Biol*, 7(1):e1000015, 2009.

Why Doesn't Everyone Do Everything? the Cooperation Conundrum of Engineered Consortia for the Catabolism of Phosphotriester Pesticides

- Conference: Synthetic Biology Engineering Evolution Design SEED
- Year: 2017
- Proceeding: 2017 Synthetic Biology: Engineering, Evolution & Design (SEED)
- Group: Poster Session
- Session:

  Confirmed Posters

Authors:
Kehila, D., UBC
Tokuriki, N., University of British Columbia
Recent studies into the interplay between the microbiome and host have extensively implicated short-chain fatty acid (SCFA) fermentation products as contributors to both intra and extra intestinal health. In the context of the gut, butyrate (a four carbon SCFA) serves as the primary energy source for epithelial colonocytes, drives regulatory T cell differentiation and is therefore essential for the maintenance of colonic mucosal health. On a systemic level, butyrate may influence insulin sensitivity, cholesterol synthesis and autoimmunity. Typically, butyrate is produced by the action of slow-growing Gram positive organisms â“ often Clostridial in origin â“ during the fermentation of cellulose and complex carbohydrates. To counteract the depletion of butyrate in the gut, we iteratively engineered *Escherichia coli* Nissle 1917 to produce butyrate. The new butyrate-producing Nissle was capable of millimolar level production of butyrate comparable with the levels produced by enteric Clostridial species. The circuitry was placed under anaerobic control to support induction upon entry into the gastrointestinal track. Finally, proof-of-concept *in vivo* experimentation showed production of butyrate in the mouse gut. This engineered strain may be useful for the treatment of a variety of gastrointestinal and systemic conditions associated with SCFA deficiency.

AIChE
120 Wall Street, FL 23
New York, NY 10005-4020

CURRAN ASSOCIATES INC.
proceedings.com

ISBN 978-1-5108-5673-8